Pooled Cross-Sectional and Time Series Data Analysis

Vol. 1: The Generalized Jackknife Statistic, *H. L. Gray and W. R. Schucany*
Vol. 2: Multivariate Analysis, *Anant M. Kshirsagar*
Vol. 3: Statistics and Society, *Walter T. Federer*
Vol. 4: Multivariate Analysis: A Selected and Abstracted Bibliography, 1957-1972, *Kocherlakota Subrahmaniam and Kathleen Subrahmaniam* (out of print)
Vol. 5: Design of Experiments: A Realistic Approach, *Virgil L. Anderson and Robert A. McLean*
Vol. 6: Statistical and Mathematical Aspects of Pollution Problems, *John W. Pratt*
Vol. 7: Introduction to Probability and Statistics (in two parts), Part I: Probability; Part II: Statistics, *Narayan C. Giri*
Vol. 8: Statistical Theory of the Analysis of Experimental Designs, *J. Ogawa*
Vol. 9: Statistical Techniques in Simulation (in two parts), *Jack P. C. Kleijnen*
Vol. 10: Data Quality Control and Editing, *Joseph I. Naus* (out of print)
Vol. 11: Cost of Living Index Numbers: Practice, Precision, and Theory, *Kali S. Banerjee*
Vol. 12: Weighing Designs: For Chemistry, Medicine, Economics, Operations Research, Statistics, *Kali S. Banerjee*
Vol. 13: The Search for Oil: Some Statistical Methods and Techniques, *edited by D. B. Owen*
Vol. 14: Sample Size Choice: Charts for Experiments with Linear Models, *Robert E. Odeh and Martin Fox*
Vol. 15: Statistical Methods for Engineers and Scientists, *Robert M. Bethea, Benjamin S. Duran, and Thomas L. Boullion*
Vol. 16: Statistical Quality Control Methods, *Irving W. Burr*
Vol. 17: On the History of Statistics and Probability, *edited by D. B. Owen*
Vol. 18: Econometrics, *Peter Schmidt*
Vol. 19: Sufficient Statistics: Selected Contributions, *Vasant S. Huzurbazar (edited by Anant M. Kshirsagar)*
Vol. 20: Handbook of Statistical Distributions, *Jagdish K. Patel, C. H. Kapadia, and D. B. Owen*
Vol. 21: Case Studies in Sample Design, *A. C. Rosander*
Vol. 22: Pocket Book of Statistical Tables, *compiled by R. E. Odeh, D. B. Owen, Z. W. Birnbaum, and L. Fisher*
Vol. 23: The Information in Contingency Tables, *D. V. Gokhale and Solomon Kullback*

Vol. 24: Statistical Analysis of Reliability and Life-Testing Models: Theory and Methods, *Lee J. Bain*
Vol. 25: Elementary Statistical Quality Control, *Irving W. Burr*
Vol. 26: An Introduction to Probability and Statistics Using BASIC, *Richard A. Groeneveld*
Vol. 27: Basic Applied Statistics, *B. L. Raktoe and J. J. Hubert*
Vol. 28: A Primer in Probability, *Kathleen Subrahmaniam*
Vol. 29: Random Processes: A First Look, *R. Syski*
Vol. 30: Regression Methods: A Tool for Data Analysis, *Rudolf J. Freund and Paul D. Minton*
Vol. 31: Randomization Tests, *Eugene S. Edgington*
Vol. 32: Tables for Normal Tolerance Limits, Sampling Plans, and Screening, *Robert E. Odeh and D. B. Owen*
Vol. 33: Statistical Computing, *William J. Kennedy, Jr. and James E. Gentle*
Vol. 34: Regression Analysis and Its Application: A Data-Oriented Approach, *Richard F. Gunst and Robert L. Mason*
Vol. 35: Scientific Strategies to Save Your Life, *I. D. J. Bross*
Vol. 36: Statistics in the Pharmaceutical Industry, *edited by C. Ralph Buncher and Jia-Yeong Tsay*
Vol. 37: Sampling from a Finite Population, *J. Hajek*
Vol. 38: Statistical Modeling Techniques, *S. S. Shapiro*
Vol. 39: Statistical Theory and Inference in Research, *T. A. Bancroft and C.-P. Han*
Vol. 40: Handbook of the Normal Distribution, *Jagdish K. Patel and Campbell B. Read*
Vol. 41: Recent Advances in Regression Methods, *Hrishikesh D. Vinod and Aman Ullah*
Vol. 42: Acceptance Sampling in Quality Control, *Edward G. Schilling*
Vol. 43: The Randomized Clinical Trial and Therapeutic Decisions, *edited by Niels Tygstrup, John M. Lachin, and Erik Juhl*
Vol. 44: Regression Analysis of Survival Data in Cancer Chemotherapy, *Walter H. Carter, Jr., Galen L. Wampler, and Donald M. Stablein*
Vol. 45: A Course in Linear Models, *Anant M. Kshirsagar*
Vol. 46: Clinical Trials: Issues and Approaches, *edited by Stanley H. Shapiro and Thomas H. Louis*
Vol. 47: Statistical Analysis of DNA Sequence Data, *edited by B. S. Weir*
Vol. 48: Nonlinear Regression Modeling: A Unified Practical Approach, *David A. Ratkowsky*
Vol. 49: Attribute Sampling Plans, Tables of Tests and Confidence Limits for Proportions, *Robert E. Odeh and D. B. Owen*
Vol. 50: Experimental Design, Statistical Models, and Genetic Statistics, *edited by Klaus Hinkelmann*
Vol. 51: Statistical Methods for Cancer Studies, *edited by Richard G. Cornell*
Vol. 52: Practical Statistical Sampling for Auditors, *Arthur J. Wilburn*
Vol. 53: Statistical Signal Processing, *edited by Edward J. Wegman and James G. Smith*
Vol. 54: Self-Organizing Methods in Modeling: GMDH Type Algorithms, *edited by Stanley J. Farlow*
Vol. 55: Applied Factorial and Fractional Designs, *Robert A. McLean and Virgil L. Anderson*
Vol. 56: Design of Experiments: Ranking and Selection, *edited by Thomas J. Santner and Ajit C. Tamhane*
Vol. 57: Statistical Methods for Engineers and Scientists. Second Edition, Revised and Expanded, *Robert M. Bethea, Benjamin S. Duran, and Thomas L. Boullion*
Vol. 58: Ensemble Modeling: Inference from Small-Scale Properties to Large-Scale Systems, *Alan E. Gelfand and Crayton C. Walker*
Vol. 59: Computer Modeling for Business and Industry, *Bruce L. Bowerman and Richard T. O'Connell*
Vol. 60: Bayesian Analysis of Linear Models, *Lyle D. Broemeling*

Vol. 61: Methodological Issues for Health Care Surveys, *Brenda Cox and Steven Cohen*

Vol. 62: Applied Regression Analysis and Experimental Design, *Richard J. Brook and Gregory C. Arnold*

Vol. 63: Statpal: A Statistical Package for Microcomputers – PC-DOS Version for the IBM PC and Compatibles, *Bruce J. Chalmer and David G. Whitmore*

Vol. 64: Statpal: A Statistical Package for Microcomputers – Apple Version for the II, II+, and IIe, *David G. Whitmore and Bruce J. Chalmer*

Vol. 65: Nonparametric Statistical Inference, Second Edition, Revised and Expanded, *Jean Dickinson Gibbons*

Vol. 66: Design and Analysis of Experiments, *Roger G. Petersen*

Vol. 67: Statistical Methods for Pharmaceutical Research Planning, *Sten W. Bergman and John C. Gittins*

Vol. 68: Goodness-of-Fit Techniques, *edited by Ralph B. D'Agostino and Michael A. Stephens*

Vol. 69: Statistical Methods in Discrimination Litigation, *edited by D. H. Kaye and Mikel Aickin*

Vol. 70: Truncated and Censored Samples from Normal Populations, *Helmut Schneider*

Vol. 71: Robust Inference, *M. L. Tiku, W. Y. Tan, and N. Balakrishnan*

Vol. 72: Statistical Image Processing and Graphics, *edited by Edward J. Wegman and Douglas J. DePriest*

Vol. 73: Assignment Methods in Combinatorial Data Analysis, *Lawrence J. Hubert*

Vol. 74: Econometrics and Structural Change, *Lyle D. Broemeling and Hiroki Tsurumi*

Vol. 75: Multivariate Interpretation of Clinical Laboratory Data, *Adelin Albert and Eugene K. Harris*

Vol. 76: Statistical Tools for Simulation Practitioners, *Jack P. C. Kleijnen*

Vol. 77: Randomization Tests, Second Edition, *Eugene S. Edgington*

Vol. 78: A Folio of Distributions: A Collection of Theoretical Quantile-Quantile Plots, *Edward B. Fowlkes*

Vol. 79: Applied Categorical Data Analysis, *Daniel H. Freeman, Jr.*

Vol. 80: Seemingly Unrelated Regression Equations Models : Estimation and Inference, *Virendra K. Srivastava and David E. A. Giles*

Vol. 81: Response Surfaces: Designs and Analyses, *Andre I. Khuri and John A. Cornell*

Vol. 82: Nonlinear Parameter Estimation: An Integrated System in BASIC, *John C. Nash and Mary Walker-Smith*

Vol. 83: Cancer Modeling, *edited by James R. Thompson and Barry W. Brown*

Vol. 84: Mixture Models: Inference and Applications to Clustering, *Geoffrey J. McLachlan and Kaye E. Basford*

Vol. 85: Randomized Response: Theory and Techniques, *Arijit Chaudhuri and Rahul Mukerjee*

Vol. 86: Biopharmaceutical Statistics for Drug Development, *edited by Karl E. Peace*

Vol. 87: Parts per Million Values for Estimating Quality Levels, *Robert E. Odeh and D. B. Owen*

Vol. 88: Lognormal Distributions: Theory and Applications, *edited by Edwin L. Crow and Kunio Shimizu*

Vol. 89: Properties of Estimators for the Gamma Distribution, *K. O. Bowman and L. R. Shenton*

Vol. 90: Spline Smoothing and Nonparametric Regression, *Randall L. Eubank*

Vol. 91: Linear Least Squares Computations, *R. W. Farebrother*

Vol. 92: Exploring Statistics, *Damaraju Raghavarao*

Vol. 93: Applied Time Series Analysis for Business and Economic Forecasting, *Sufi M. Nazem*

Vol. 94: Bayesian Analysis of Time Series and Dynamic Models, *edited by James C. Spall*

Vol. 95: The Inverse Gaussian Distribution: Theory, Methodology, and Applications, *Raj S. Chhikara and J. Leroy Folks*

Vol. 96: Parameter Estimation in Reliability and Life Span Models, *A. Clifford Cohen and Betty Jones Whitten*

Vol. 97: Pooled Cross-Sectional and Time Series Data Analysis, *Terry E. Dielman*

ADDITIONAL VOLUMES IN PREPARATION

Pooled Cross-Sectional and Time Series Data Analysis

TERRY E. DIELMAN
Texas Christian University
Fort Worth, Texas

MARCEL DEKKER, INC. New York and Basel

ISBN 0-8247-7864-2

MARCEL DEKKER, INC.
270 Madison Avenue, New York, New York 10016

Current printing (last digit):
10 9 8 7 6 5 4 3 2 1

PRINTED IN THE UNITED STATES OF AMERICA

Preface

In the econometrics literature, a data base that provides a multivariate statistical history for each of a number of individual entities is called a pooled cross-sectional and time series data base. In marketing and survey literature, the terms panel data or longitudinal data are often used. In management science, a convenient term might be management data base. Such a data base provides a particularly rich environment for statistical analysis. Pooled data contain the information necessary to deal with both the intertemporal dynamics and the individuality of the entities being investigated.

This book reviews methods for estimating multivariate relationships particular to each individual entity in the data base and for summarizing these relationships for a number of individuals. Methodologies examined include classical pooling, error components, analysis of covariance, seemingly unrelated regression, and random coefficient regression. In each case the model is presented with assumptions necessary for estimation and inference, examples are given, the limitations of the model are discussed, and available computer software is mentioned. The orientation of the book is applied, with emphasis on analysis of data. Properties of various estimators are discussed, but proofs are limited to those not readily available in the literature.

Considerations in choosing an appropriate model are provided. Small-sample simulation results available in the literature are reviewed and some additional results obtained by the author are presented to aid in model choice.

This book provides a more comprehensive review of methodologies for analyzing pooled cross-sectional and time series data than is otherwise available. Presentation of examples and available computer software should make the book especially appealing to applied statisticians and data analysts. It should also be useful as a supplementary text in advanced econometrics or regression analysis courses as well as a reference for academicians teaching or doing research in areas related to econometrics or linear models.

In writing this book I have attempted to provide a comprehensive list of references for the various methodologies discussed. However, I am sure there are omissions. I apologize to the authors for these and would be interested in being notified of any important omissions.

Acknowledgments are due to the reader at Marcel Dekker, Inc., whose detailed comments aided in improving this manuscript. Also, thanks to Professor Roger Wright for first directing me to this area of study, Professor Roger Pfaffenberger for his comments on an earlier draft, and Linda Tarvin and Marty Burkhard for patience in typing the manuscript. Of course, any errors in the manuscript are the responsibility of the author.

This research was partially funded by a summer research grant from the M. J. Neeley School of Business, Texas Christian University.

Terry E. Dielman

Contents

1
Introduction

1.1 POOLED CROSS-SECTIONAL AND TIME SERIES DATA DEFINED

Statistical methods can be characterized according to the type of data to which they are applied. The field of survey statistics usually deals with cross-sectional data describing each of many different individuals or units at a single point in time. Econometrics commonly uses time series data describing a single entity-usually an economy or market. Although other types of statistical data exist, such as geographical data or data produced by controlled experiments, most researchers currently use methods of analysis developed for either cross-sectional or time series data.

Data analysts often encounter yet another type of data that will be called pooled data. Pooled data means any data base describing each of a number of individuals across a sequence of time periods. For example, a financial market data base may contain the monthly rates of return for a sample of securities (Dielman, Nantell, and Wright 1980) whereas marketing data may describe sales of certain brands of products through time (Beckwith 1972); in developing a demand function one might measure demand for gasoline through time for different states (Mehta, Narasimham, and Swamy 1978), demand for automobiles over time by the size of the car (Carlson 1978), or hourly demand for electricity by individual households (Granger et al. 1979); to construct a yield function for a pine tree plantation, volume on individual plots over time could be measured (Ferguson and Leech 1978); medical research might develop a data base of drug dosage and subsequent blood-sugar-level measurements for a number of patients at periodic intervals (Sheiner, Rosenberg, and Melmon 1972). A pooled data base thus blends characteristics of both cross-sectional and time series data. Like cross-sectional data, it describes each of a number of individuals. Like time series data, it describes each single individual through time. Pooled data are important to the analyst because they contain the information necessary to deal with both the intertemporal dynamics and the individuality of the entities being investigated.

1.2 SPECIAL PROBLEMS ASSOCIATED WITH ANALYZING POOLED DATA

To realize the potential value of the information contained in a pooled data base, the analyst needs a statistical methodology encompassing the following functions:

1. Study of the performance of a single particular individual described in the data base, and relating this performance to appropriate explanatory variables. For example, in establishing flexible budgetary goals for an automobile dealer, the analyst might need to relate the dealer's sales history to level of service, promotional expenditures, and competitive environment. Generally, multiple regression analysis can be used to analyze the available time series data describing any particular individual.

2. Summarization of individual relationships describing each of a sample of individuals, and drawing generalized inferences from these summary statistics. For example, an automotive firm may want to determine the average effect of promotional expenditure on sales for a group of its dealers, to examine the range of such effects, and to test hypotheses about the distribution of effects within the population of all its dealers.

These are special problems associated with the estimation of regression equations describing the behavior of individuals over time. In a single time series equation we often encounter the problem of autocorrelated errors. In a single cross-sectional regression equation we might have problems with nonconstant error variance. When combining cross-sectional and time series data we may find both problems occurring simultaneously. In addition, we may find that cross-sectional disturbances for different individuals at the same point in time may be correlated. Such contemporaneous correlation is an added problem specific to the analysis of pooled data.

The behavior of individuals themselves complicates the specification of the relationship to be estimated when pooled data are used. When estimating a regression equation for cross-sectional data it is often assumed that the equation is valid for each individual in the sample. In the automotive dealer example, the automotive firm might assume that for given values of promotional expenditure, two individual dealers would respond in exactly the same fashion in terms of sales. A more reasonable assumption might be that individual dealer's responses might differ, even if promotional expenditure were held constant.

As Klein (1953, pp. 211-225) points out, when we analyze data pertaining to a number of individuals, one possibility is that the underlying relationship between an independent and dependent variable may be different for different individuals. The use of pooled data allows us the possibility of modeling these differences and estimating individual response coefficients or average responses for the group.

When estimating a time series equation for a single cross-sectional unit we may find that a regression coefficient changes over time (either systematically or randomly). This can also be the case when pooling time series for several individuals. As when cross-sectional relationships between a dependent and an independent variable differ between individuals, we can have a relationship between a dependent and an independent variable that changes over time or we can have both changes occurring simultaneously. Under certain assumptions we can model such processes and estimate coefficients pertaining to individuals or estimate averages for the group.

1.3 PRELIMINARIES: GENERALIZED LEAST SQUARES ESTIMATION AND NOTATION

Consider the equation

$$Y = X\beta + \varepsilon \tag{1.1}$$

where Y is a vector of observations on the dependent variable, X is a matrix of observations on the independent or explanatory variables, β is a vector of regression coefficients to be estimated and ε is a vector of disturbances.

Assume that $E(\varepsilon) = 0$ and $E(\varepsilon\varepsilon') = \Omega$. The generalized least squares (GLS) estimator of β is given by

$$\tilde{\beta} = (X'\Omega^{-1}X)^{-1}X'\Omega^{-1}Y \tag{1.2}$$

Throughout this book the "~" notation will be used to indicate the GLS estimator assuming the disturbance variance-covariance matrix, Ω, is known. Also assume that X is of full rank and is fixed in repeated samples. Then $\tilde{\beta}$ will be efficient among the class of all linear, unbiased estimators of β (see Fomby, Hill and Johnson (1984, pp. 17)). The estimator $\tilde{\beta}$ is referred to as the generalized least squares estimator or Aitken estimator. Since the variance-covariance matrix, Ω, is typically unknown in practice, it is often not possible to use the GLS estimator.

When Ω can be consistently estimated by, say, $\hat{\Omega}$, the following estimator can be used:

$$\hat{\beta} = (X'\hat{\Omega}^{-1}X)^{-1}X'\hat{\Omega}^{-1}Y \tag{1.3}$$

The estimator $\hat{\beta}$ is referred to as the feasible GLS or feasible Aitken estimator. The '^' notation will be used to denote the feasible GLS estimator.

Fomby, Hill and Johnson (1984, pp. 148-154) summarize the properties of the feasible GLS estimator. Since many of the estimators constructed throughout this text are feasible GLS estimators, the properties presented in Fomby et al., will be outlined here.

The following assumptions are made for the model in equation (1.1):

Assumption 1.1: $\hat{\Omega}$ is of full rank so that $\hat{\Omega}^{-1}$ exists.

Assumption 1.2: The number of parameters in Ω plus the number of regression coefficients to be estimated must be smaller than the sample size.

Assumption 1.3: Consistent estimates of the parameters in Ω must exist.

Assumption 1.4: There exists a matrix P such that $PP' = \Omega^{-1}$ and the elements of $P'\varepsilon$ are independently and identically distributed.

Assumption 1.5: $\operatorname{plim} \dfrac{X'\hat{\Omega}^{-1}X}{T} = \operatorname{plim} \dfrac{X'\Omega^{-1}X}{T} = Q$ where Q is finite and nonsingular.

Assumption 1.6: $\operatorname{plim} \dfrac{X'\hat{\Omega}^{-1}e}{T} = \operatorname{plim} \dfrac{X'\Omega^{-1}e}{T} = 0.$

Assumption 1.7: $\operatorname{plim} \dfrac{e'\hat{\Omega}^{-1}e}{T} = \operatorname{plim} \dfrac{e'\Omega^{-1}e}{T} = \sigma^2.$

Then $\hat{\beta}$ will be a consistent estimator and $\sqrt{T}(\hat{\beta} - \beta)$ will be asymptotically distributed as

$$N(0, \sigma^2 Q^{-1})$$

Fomby et al. also provide results that will hold concerning variance estimates and hypothesis tests.

1.4 HISTORICAL PERSPECTIVE

One of the earliest methods of treating time series and cross-sectional data in the same analysis involved using cross-sectional data to estimate some of the model parameters, then introducing these estimates into a time series regression equation as though they were known with certainty. The time series regression then provides estimates of the additional parameters conditional upon the estimates obtained from the cross-sectional data. This approach was used by Klein (1953, pp. 211-225), Marschack (1943), Solow (1964), Staehle (1945), Stone (1954), Tobin (1950) and Wold and Jureen (1951). A discussion of the drawbacks of this approach is found in Kuh and Meyer (1957). Haavelmo (1947) also provides an early discussion on the relationship between estimates of the same equation using cross-sectional data and time series data, separately. In these studies, as noted, the parameter estimates obtained from the cross-sectional regressions are introduced into the time series equations as though they are error free. The information being introduced is obviously not without error, however, since it is derived from the cross-sectional parameter estimates.

Recognizing the problem with such an approach, Durbin (1953), Theil and Goldberger (1961) and Theil (1963) suggested methods of introducing extraneous information in the form of estimates of some of the coefficients into a regression model.

Theil and Goldberger (1961), for example, proposed a method termed the method of mixed estimation in which they suggested using cross-sectional estimates in the time series regressions but without the assumption that the estimates are free of error. Consider the time series regression equation

$$Y = X\beta + \varepsilon \tag{1.4}$$

where Y is a $T \times 1$ vector of observations on the dependent variable, X is a $T \times K$ matrix of observations on the explanatory variables, β is a $K \times 1$ vector of unknown parameters to be estimated, and ε is a $T \times 1$ vector of random disturbances. Assume that $E(\varepsilon) = 0$ and $E(\varepsilon\varepsilon') = V_1$.

From a cross-sectional data base suppose an estimate, b, of β is obtained which can be written

$$b = R\beta + v \tag{1.5}$$

with $E(v) = 0$, $E(vv') = V_2$ since the estimate is related to β but not without error, and $E(\varepsilon v) = 0$. Combining equations (1.4) and (1.5)

$$\begin{bmatrix} Y \\ b \end{bmatrix} = \begin{bmatrix} X \\ R \end{bmatrix} \beta + \begin{bmatrix} \varepsilon \\ v \end{bmatrix} \tag{1.6}$$

where

$$E \begin{bmatrix} \varepsilon \\ v \end{bmatrix} [\varepsilon' \; v'] = \begin{bmatrix} V_1 & 0 \\ 0 & V_2 \end{bmatrix}$$

Equation (1.6) can be estimated by generalized least squares (GLS) as follows:

$$\tilde{\beta} = (X' V_1^{-1} X + R' V_2^{-1} R)^{-1} (X' V_1^{-1} Y + R' V_2^{-1} b) \tag{1.7}$$

Young (1972) applied the Theil-Goldberger method while properties of this or similar estimators were examined by several authors, including Nagar and Kakwani (1964), Swamy and Mehta (1969, 1977b), Mehta and Swamy (1970, 1974), Kakwani (1968), Suzuki (1964), Ohtani and Honda (1984), Yancey, Bock, and Judge (1972), Theil (1974) and Paulus (1975). Others studying the use of estimates from cross-sectional regressions in time series regressions include Chetty (1968), Desai (1974), Ghalai (1977), Izan (1980), Maddala (1971a) and Mikhail (1975).

The mixed estimation procedure involves some asymmetry in the use of the time series and cross sectional samples. Previously, an estimate from a cross-sectional data base was incorporated as extraneous information into a time series regression. The role of the two samples could just as easily have been reversed, however. To overcome this asymmetry, Theil and Goldberger (1961) also suggest a balanced method of using the statistical information which involves an extension of their mixed estimation approach. Although this procedure allows the use of the time series and cross section data together in a single analysis the technique applies primarily to situations where there is one cross-sectional relationship and one time series relationship, each of which have certain parameters in common.

An alternative to using cross-sectional estimates as extraneous information in time series regressions was to aggregate the data for N individuals and to estimate a single time series regression equation. The time series equation for individual i can be written as

$$Y_i = X_i\beta + \varepsilon_i \tag{1.8}$$

where Y_i is a $T \times 1$ vector of observations on the ith individual, X_i is a $T \times K$ matrix of observations on the ith individual, β is a $K \times 1$ vector of parameters to be estimated, and ε_i is a $T \times 1$ vector of disturbances.

Assume that $E(\varepsilon_i) = 0$ and $E(\varepsilon_i \varepsilon_i') = \sigma^2 I$. There are N such time series equations corresponding to the N individuals in the sample. First, aggregate the data for all individuals:

$$\overline{Y} = \frac{1}{N}\sum_{i=1}^{N} Y_i \tag{1.9}$$

$$\overline{X} = \frac{1}{N}\sum_{i=1}^{N} X_i \tag{1.10}$$

Then, the aggregate time series regression can be written

$$\overline{Y} = \overline{X}\beta + \frac{1}{N}\sum_{i=1}^{N} \varepsilon_i \tag{1.11}$$

and the <u>ordinary least squares</u> (OLS) estimator of β is

$$\hat{\beta}_A = (\overline{X}'\,\overline{X})^{-1}\,\overline{X}'\,\overline{Y} \tag{1.12}$$

Aggregation assumes that the coefficient vector β is the same for each individual in the sample. If this is true the estimator

$\hat{\beta}_A$ will be unbiased. If the individual coefficient vectors differ, OLS estimation may result in what is termed aggregation bias (see Theil 1971, pp. 556-562). Zellner (1969) pointed out, however, that differences in the coefficients may not lead to aggregation bias if the individual coefficient vectors can be written as

$$\beta_i = \bar{\beta} + v_i \tag{1.13}$$

where v_i is a random component with expected value zero. In this case the estimator $\hat{\beta}_A$ may still be unbiased. [See also Akkina (1974).] Thus, if it is assumed that the coefficient vectors differ between individuals, that these differences are random, and that it is the mean or expected value of the coefficients that is to be estimated, then OLS applied to the aggregate data may yield an unbiased estimator. See Green (1964), Kuh (1974) and Theil (1954) for further discussions of aggregation.

Aggregation is usually an attempt to develop a group relation or macrorelation from a number of individual relations or microrelations. Welsch and Kuh (1976) pointed out that aggregation is not particularly appealing when nearly all of the microdata are available. Econometricians are often obliged to develop aggregate models, however, when microdata are unavailable or limited. For examples, see Atkinson and Mairesse (1978), Eisner (1978), and Oudiz (1978).

1.5 WHAT IS ATTEMPTED IN THIS BOOK?

This book reviews statistical methods for studying the individuals within a pooled data base and for summarizing individual relationships. Separate chapters are set aside for specific models which incorporate a particular set of assumptions about coefficients and/or disturbances in the models constructed. Within each chapter the assumptions of the model are set out, estimation is discussed, procedures for inference are established and examples of the use of the model are presented. Also discussed will be limitations of the model, small sample properties as determined from simulation studies (if available) and a guide to any available computer software for the technique.

In later chapters of the book some guidelines for choosing between alternative models will be presented and suggestions will be given for those areas in which further research is necessary.

For other surveys on analyzing pooled cross-sectional and time series data see Berk et al. (1979), Dielman (1983), Fanfani

(1975), Griffiths (1974), Hall (1978), Hannan and Young (1977), Swamy (1971) and Hsiao (1985, 1986). Also, many econometrics texts contain chapters on pooled data. For example, see Judge et al. (1985, Chapters 12 and 13), Pindyck and Rubinfeld (1976, Chapter 7) or Kmenta (1971, Chapter 12).

2

Separate Regressions and Classical Pooling

2.1 SEPARATE REGRESSIONS

When the performance of one individual from the pooled database is of interest, separate equation regressions can be estimated for each individual unit. Each relationship is written as

$$Y_i = X_i\beta_i + \varepsilon_i \tag{2.1}$$

for i = 1, ... , N where Y_i is a T × 1 matrix of time series observations on the dependent variable, X_i is a T × K matrix of observations on the independent variables, β_i is a K × 1 vector of parameters to be estimated, and ε_i is a T × 1 vector of disturbances. If each X_i has full column rank, then the ordinary least squares (OLS) estimator of β_i is given by

$$\hat{\beta}_i = (X_i'X_i)^{-1}X_i'Y_i \tag{2.2}$$

In order for $\hat{\beta}$ to be a best linear (in Y_i) unbiased estimator (BLUE) of β_i, the following assumtions must hold:

Assumption 2.1: $E(\varepsilon_i) = 0$.

Assumption 2.2: $E(\varepsilon_i\varepsilon_i') = \sigma_i^2 I_T$.

Assumption 2.3: $E(\varepsilon_i\varepsilon_j') = 0$, for $i \neq j$.

Assumption 2.4: X_i is a fixed matrix (in repeated samples).

These conditions are sufficient but not necessary for the optimality of the OLS estimator (see Rao and Mitra, 1971, Ch. 8). When OLS is not optimal, estimation can still proceed equation by equation in many cases. For example, if Assumption 2.2 is violated and disturbances are either serially correlated or heteroskedastic, generalized least squares (GLS), applied to each equation separately, will provide relatively more efficient estimates than OLS.

For an example, see Gendreau and Humphrey (1980).

2.2 CLASSICAL POOLING

With separate equation regressions, parameter estimates are obtained for each individual unit in the data base. Now suppose it is necessary to summarize individual relationships and, from the summary statistics, to draw inferences about certain population parameters. Alternatively, this process might be viewed as building a single model to describe the entire group of individuals rather than building a separate model for each.

One approach to this process assumes that Assumptions 2.1 to 2.4 hold and adds the additional assumption that the population regression coefficients are equal for all individuals. This can be stated as Assumption 2.5:

Assumption 2.5: $\beta_1 = \beta_2 = \cdots = \beta_N = \bar{\beta}$.

We are assuming that the individuals in our database are drawn from a population with a common regression parameter vector $\bar{\beta}$. In this case the observations for each individual can be pooled and a single regression performed to obtain a more efficient estimator of $\bar{\beta}$. The equation system is now written as

$$Y = Z\bar{\beta} + \varepsilon \qquad (2.3)$$

where

$$Y = \begin{bmatrix} Y_1 \\ Y_2 \\ \cdot \\ \cdot \\ \cdot \\ Y_N \end{bmatrix}_{NT \times 1} \qquad Z = \begin{bmatrix} X_1 \\ X_2 \\ \cdot \\ \cdot \\ \cdot \\ X_N \end{bmatrix}_{NT \times K} \qquad \varepsilon = \begin{bmatrix} \varepsilon_1 \\ \varepsilon_2 \\ \cdot \\ \cdot \\ \cdot \\ \varepsilon_N \end{bmatrix}_{NT \times 1}$$

and $\bar{\beta}$ is a $K \times 1$ vector of coefficients to be estimated.

If Assumption 2.2 is replaced by:

Assumption 2.2A: $E(\varepsilon_i \varepsilon_i') = \sigma^2 I_T$,

that is, if the error variance can be assumed equal for each individual, then $\bar{\beta}$ is estimated efficiently and without bias by

$$\hat{\bar{\beta}}_{CP1} = (Z'Z)^{-1}Z'Y \tag{2.5}$$

This estimator has been termed the <u>classical pooling</u> (CP) estimator. Assumption 2.6 is a very restrictive one in terms of how we would expect disturbances for separate individuals to behave. For example, it would be more likely that Assumption 2.2 would hold with different disturbance variance for each individual. At the least, it should be possible to allow for different variances. The CP estimator under Assumption 2.2 would be

$$\tilde{\bar{\beta}}_{CP2} = (Z'\Omega^{-1}Z)^{-1}Z'\Omega^{-1}Y \tag{2.6}$$

$$\text{where } \Omega = \begin{bmatrix} \sigma_1^2 I_T & 0 & \cdots & 0 \\ 0 & \sigma_2^2 I_T & \cdots & 0 \\ \vdots & \vdots & \ddots & \vdots \\ 0 & 0 & \cdots & \sigma_N^2 I_T \end{bmatrix} \tag{2.7}$$

The unknown parameters σ_i^2 can be consistently estimated by

$$s_i^2 = \frac{1}{T-K} \sum_{t=1}^{T} \hat{\varepsilon}_{it}^2 \qquad i = 1, \ldots, N \tag{2.8}$$

where $\hat{\varepsilon}_{it}$ are the residuals obtained from applying OLS to equation (2.1).

Kmenta (1971, pp. 508-514) suggests even more general assumptions in the classical pooling model. Combining the assumption usually made when dealing with cross-sectional data, that the error variance will differ between cross sections, with that made when using time series data, that time series residuals will be autocorrelated, he arrives at his cross-sectionally heteroskedastic and timewise autoregressive model. The assumptions 2.1 to 2.5 are maintained but, in addition, the following assumption is made:

Assumption 2.6: $\varepsilon_{it} = \rho_i \varepsilon_{i,t-1} + u_{it}$

where

ε_{it} is the tth element of the vector ε_i

$$E(u_{it}) = 0$$

$$E(u_{it}^2) = \sigma_{ui}^2$$

$$E(\varepsilon_{io}^2) = \frac{\sigma_{ui}^2}{1 - \rho_i^2}$$

and $E(\varepsilon_{i,t-1}\, u_{jt}) = 0$ for all i and j.

The u_{it} are serially uncorrelated disturbances which are also uncorrelated with the original disturbances, $\varepsilon_{i,t-1}$ from the previous time period. The assumption concerning $E(\varepsilon_{io}^2)$ provides an initial condition for the disturbance generating process.

The CP estimator under these assumptions would be

$$\tilde{\beta}_{CP3} = (Z'\Omega^{-1}Z)^{-1}\, Z'\Omega^{-1}Y \tag{2.9}$$

where

$$\Omega = \begin{bmatrix} \sigma_1^2 P_1 & 0 & \cdots & 0 \\ 0 & \sigma_2^2 P_2 & \cdots & 0 \\ \vdots & \vdots & \ddots & \vdots \\ 0 & 0 & & \sigma_N^2 P_N \end{bmatrix} \tag{2.10}$$

and

$$P_i = \begin{bmatrix} 1 & \rho_i & \rho_i^2 & \cdots & \rho_i^{T-1} \\ \rho_i & 1 & \rho_i & \cdots & \rho_i^{T-2} \\ \vdots & \vdots & \vdots & & \vdots \\ \rho_i^{T-1} & \rho_i^{T-2} & \rho_i^{T-3} & \cdots & 1 \end{bmatrix} \tag{2.11}$$

Estimators of the ρ_i can be found using

$$\hat{\rho}_i = \frac{\Sigma_{t=2}^{T} \hat{\varepsilon}_{it} \hat{\varepsilon}_{i,t-1}}{\Sigma_{t=2}^{T} \hat{\varepsilon}_{i,t-1}} \tag{2.12}$$

Next the original data must be transformed so that a relationship is obtained in which the disturbances are serially uncorrelated. The equation of such a relationship can be written

$$Y_i^* = X_i^* \bar{\beta} + u_i \tag{2.13}$$

The transformation to obtain the elements of Y_i^* and X_i^* are

$$\begin{aligned} y_{it}^* &= y_{it} - \hat{\rho}_i y_{i,t-1} \\ x_{itk}^* &= x_{itk} - \hat{\rho}_i x_{i,t-1,k} \end{aligned} \tag{2.14}$$

where y_{it} and y_{it}^* indicate the tth elements of the vectors Y_i and Y_i^*, respectively, and x_{itk} and x_{itk}^* represent the tth values of the kth independent variable from the matrices X_i and X_i^*, respectively. The transformations (2.14) apply to observations t = 2,..., T. The first observation can be transformed as

$$\begin{aligned} y_{i1}^* &= (1 - \hat{\rho}_i^2)^{\frac{1}{2}} y_{i1} \\ x_{i1k}^* &= (1 - \hat{\rho}_i^2)^{\frac{1}{2}} x_{i1k} \end{aligned} \tag{2.15}$$

Often the first observation is merely discarded rather than transformed. The justification for this is that, asymptotically, the loss of one observation will not matter. Empirical work in several contexts has shown that the inclusion of the differentially transformed first observation can provide substantial gains in efficiency, however. See, for example, Park and Mitchell (1980) and Maeshiro (1979). Therefore, throughout this text the transformed first observation will be included whenever corrections for first-order autocorrelation are made.

Applying OLS to the equations (2.13) provides us with vectors of residuals, $\hat{u}_i$, that can be used to estimate the disturbance variance,

$$\hat{\sigma}_{ui}^2 = \frac{1}{T - K} \sum_{t=1}^{T} \hat{u}_{it}^2 \tag{2.16}$$

Then σ_i^2 can be estimated by

$$\hat{\sigma}_i^2 = \frac{\hat{\sigma}_{ui}^2}{1 - \hat{\rho}_i^2} \tag{2.17}$$

Note that $\hat{\rho}_i$ and $\hat{\sigma}_{ui}^2$ are consistent estimators of ρ_i and σ_{ui}^2 respectively. It follows that $\hat{\sigma}_i^2$ will be a consistent estimator of σ_i^2.

The estimator $\hat{\bar{\beta}}_{CP3}$ can now be computed by substituting the estimators $\hat{\rho}_i$ and $\hat{\sigma}_i^2$ into Ω.

$$\hat{\bar{\beta}}_{CP3} = (Z'\hat{\Omega}^{-1}Z)^{-1} Z'\hat{\Omega}^{-1}Y \tag{2.18}$$

Computationally, a simple way to obtain $\hat{\bar{\beta}}_{CP3}$ would be to use a double transformation of the data and then to apply OLS:

1. Transform to remove first-order autocorrelation as was done to obtain equations (2.13).
2. Divide through these equations by $\hat{\sigma}_{ui}$.
3. Then apply OLS.

In addition, Kmenta suggests relaxing the assumption that cross-sectional units are independent. That is, it is possible to allow for contemporaneous correlation of disturbances as well as first-order autocorrelation of the time series disturbances and cross-sectional heteroskedasticity.

The assumptions would be as follows:

Assumption 2.1: $E(\varepsilon_i) = 0$.

Assumption 2.2: $E(\varepsilon_i \varepsilon_j') = \sigma_i^2 I_T$.

Assumption 2.3A: $E(\varepsilon_i \varepsilon_j') = \sigma_{ij} I_T$

Assumption 2.4: X_i is a fixed matrix (in repeated samples).

Assumption 2.5: $\beta_1 = \beta_2 = \cdots = \beta_N = \bar{\beta}$.

Assumption 2.6A: $\varepsilon_{it} = \rho_i \varepsilon_{i,t-1} + u_{it}$.

where

$$E(u_{it}) = 0$$

$$E(u_{it}^2) = \sigma_{u_i}^2$$

$$E(u_{it}u_{jt}) = \sigma_{u_{ij}}$$

$$E(u_{it}u_{js}) = 0 \text{ for } t \neq s$$

$$E(\varepsilon_{io}^2) = \frac{\sigma_{u_{ij}}}{1 - \rho_i^2}$$

$$E(\varepsilon_{io}\varepsilon_{jo}) = \frac{\sigma_{u_{ij}}}{1 - \rho_i \rho_j}$$

The GLS estimator is

$$\tilde{\bar{\beta}}_{CP4} = (Z'\Omega^{-1}Z)^{-1} Z'\Omega^{-1}y \tag{2.19}$$

$$\text{where } \Omega = \begin{bmatrix} \sigma_{11}P_{11} & \sigma_{12}P_{12} & \cdots & \sigma_{1N}P_{1N} \\ \sigma_{21}P_{21} & \sigma_{22}P_{22} & \cdots & \sigma_{2N}P_{2N} \\ \vdots & \vdots & & \vdots \\ \sigma_{N1}P_{N1} & \sigma_{N2}P_{N2} & \cdots & \sigma_{NN}P_{NN} \end{bmatrix} \tag{2.20}$$

with

$$P_{ij} = \begin{bmatrix} 1 & \rho_j & \rho_j^2 & \cdots & \rho_j^{T-1} \\ \rho_i & 1 & \rho_j & \cdots & \rho_j^{T-2} \\ \cdot & & & & \cdot \\ \cdot & & & & \cdot \\ \cdot & & & & \cdot \\ \rho_i^{T-1} & \rho_i^{T-2} & \rho_i^{T-3} & \cdots & 1 \end{bmatrix} \tag{2.21}$$

To obtain estimates of the components of Ω, first estimate the ρ_i as in equation (2.12). These estimates are used to transform the original data as in equations (2.14) and (2.15) resulting in equation (2.13). Applying OLS to equation (2.13) provides estimates of the residuals, $\hat{u}_{it}$.

Now the $\sigma_{u_{ij}}{}^2$ can be estimated by

$$\hat{\sigma}_{u_{ij}}{}^2 = \frac{1}{T - K} \sum_{t=1}^{T} \hat{u}_{it}\, \hat{u}_{jt} \tag{2.22}$$

the $\sigma_{u_i}{}^2$ can be estimated as in equation (2.16), the $\sigma_{ij}{}^2$ can be estimated by

$$\hat{\sigma}_{ij}{}^2 = \frac{\hat{\sigma}_{u_{ij}}{}^2}{1 - \hat{\rho}_i \hat{\rho}_j} \tag{2.23}$$

and the $\sigma_i{}^2$ can be estimated as in equation (2.17). These estimates can be substituted into Ω to obtain the feasible Aitken estimator

$$\hat{\bar{\beta}}_{CP4} = (Z'\hat{\Omega}^{-1}Z')\, Z'\hat{\Omega}^{-1}Y \tag{2.24}$$

Note that if N > T it will not be possible to compute $\hat{\bar{\beta}}_{CP4}$. Thus, to apply the cross-sectionally correlated and timewise autoregression model, the number of time series observations should be larger than the number of cross-sections. Obviously, the large number of parameters to be estimated is a drawback for this model even when a sufficient number of observations is available.

Assuming all regression coefficients are equal across individual units may not be appropriate as pointed out by Klein (1953, pp. 211-225) and Swamy (1971, Ch. 1). It might, however, be appropriate to make the equality assumption for a subset of the coefficients and to consider different degrees of pooling.

Consider equation (2.1) but partition the X_i and β_i vectors as follows:

$$Y_i = X_{1i}\beta_{1i} + X_{2i}\beta_{2i} + \varepsilon_i \tag{2.25}$$

where X_{1i} is a $T \times K_1$ matrix of observations on K_1 of the independent variables with β_{1i} the vector of corresponding coefficients and X_{2i} is a $T \times K_2$ matrix of observations on K_2 of the independent variables with β_{2i} the vector of corresponding coefficients. Suppose that the β_{1i} can be assumed equal to $\bar{\beta}_1$ for all cross-sectional units but the β_{2i} differ. The model could then be written as

$$Y = Z\bar{\beta}_1 + X_2\beta_2 + \varepsilon \tag{2.26}$$

where

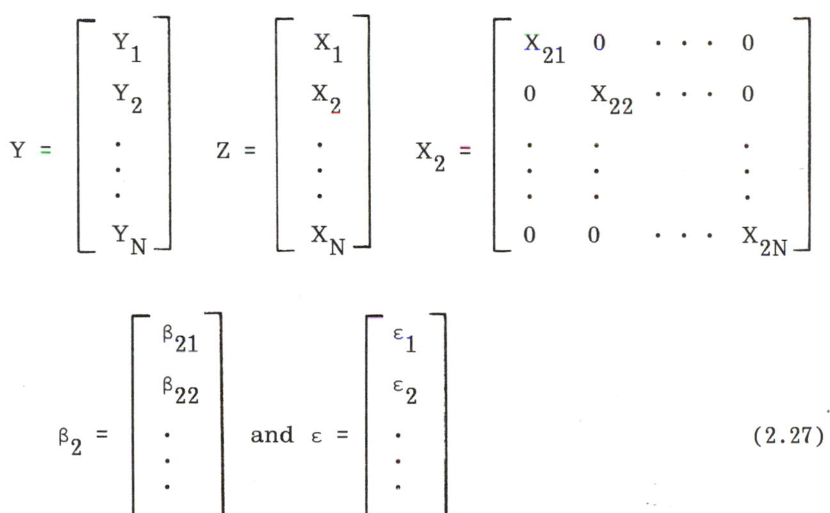

$$Y = \begin{bmatrix} Y_1 \\ Y_2 \\ \vdots \\ Y_N \end{bmatrix} \quad Z = \begin{bmatrix} X_1 \\ X_2 \\ \vdots \\ X_N \end{bmatrix} \quad X_2 = \begin{bmatrix} X_{21} & 0 & \cdots & 0 \\ 0 & X_{22} & \cdots & 0 \\ \vdots & \vdots & & \vdots \\ 0 & 0 & \cdots & X_{2N} \end{bmatrix}$$

$$\beta_2 = \begin{bmatrix} \beta_{21} \\ \beta_{22} \\ \vdots \\ \beta_{2N} \end{bmatrix} \quad \text{and} \quad \varepsilon = \begin{bmatrix} \varepsilon_1 \\ \varepsilon_2 \\ \vdots \\ \varepsilon_N \end{bmatrix} \tag{2.27}$$

Equation (2.26) can be rewritten as

$$Y = [Z \vdots X_2] \begin{bmatrix} \bar{\beta}_1 \\ \beta_2 \end{bmatrix} + \varepsilon \tag{2.28}$$

$$= W\gamma + \varepsilon \tag{2.29}$$

The vector of regression coefficients can now be estimated by either OLS if the assumptions 2.1 to 2.4 are satisfied, or GLS if one of the alternative assumptions concerning the disturbances applies.

It would also be possible to test, for example, whether the β_{2i} were equal for all individuals:

$$H_0: \beta_{21} = \beta_{22} = \cdots = \beta_{2N} = \bar{\beta}_2$$

The following test statistic could be used to conduct the test:

$$F = \frac{(SSE_2 - SSE_1)/(NK_2 + K_1 - K)}{SSE_1/(NT - K_1 - NK_2)} \tag{2.30}$$

where SSE_1 is the residual sum of squares from the unrestricted model in equation (2.26) and SSE_2 is the residual sum of squares in the restricted model. The restricted model can be written as in equation (2.3). The statistic F in (2.30) is appropriate in small samples if the disturbances, ε, are normally distributed. When ε is nonnormal, the finite sample distribution of F is unknown. However, as pointed out by Judge et al. (1985, pp. 157-159), the test statistic may still be useful even in finite samples.

The F-test could also be used to test whether Assumption 2.5 is true:

$$H_o: \beta_1 = \beta_2 = \cdots = \beta_N = \bar{\beta}.$$

This is a test of whether all coefficients are equal and the statistic is just a special case of the test statistic in (2.30).

An alternate test procedure for those cases when error variances of different equations do not differ was examined by Toro-Vizcarrondo and Wallace (1968), Wallace and Toro-Vizcarrondo (1969), Goodnight and Wallace (1972), Wallace (1972), and McElroy (1977b). See also Bass and Wittink (1975, 1978), and Brobst and Gates (1977). Swamy and Mehta (1979) analyzed the more difficult problem of pooling when error variances of different equations differ using a two-equation system. Nssah (1982) considers the same problem as Swamy and Mehta: When two equations from possibly different regimes (different coefficients and different error variances) are to be considered, how do we best make the decision of whether or not to pool the data? A literature review of other work in this area is provided by Nssah.

Examples of the use of classical pooling can be found in Blair and Kraft (1974), Feige (1974), Margolis (1982), Becker and Morey (1980), Miller and Voth (1982), Wolpin (1980) and Huang (1987).

Separate equation regressions and classical pooling can obviously be performed with any computer package with a regression routine. As more complicated error structures are introduced some additional difficulty may be encountered in obtaining estimates. However, OLS applied to suitably transformed data will produce the CP estimates described in this chapter.

Programs are available which allow for certain of the more complicated error structures. Havenner and Herman (1977) discuss one such routine to be used with the SPEAKEASY/FEDEASY processor (see Condie, 1977) which provides estimates for the cross-sectionally heteroskedastic and timewise autoregressive model of Kmenta (1971, pp. 509-512). This model is specifically discussed earlier in this chapter.

Drummond and Gallant (1977) have developed several procedures which are available from SAS (1982). They are not available in the basic SAS package however. The cross-sectionally correlated and timewise autoregressive model of Kmenta (1971, pp. 512-514) is included in their package. This model was also discussed earlier in this chapter.

2.3 EXAMPLE: INVESTMENT FUNCTION ESTIMATION

The example used in this section is taken from Boot and deWit (1960) and also examined by Vinod and Ullah (1981, pp. 258-261). The data used are reproduced in Table 2.1. Annual data are used on ten different firms for twenty years to estimate the following investment function:

$$Y_{it} = \beta_{0i} + \beta_{1i}X_{1i,t-1} + \beta_{2i}X_{2i,t-1} + \varepsilon_{it} \qquad (2.31)$$

where

TABLE 2.1 Investment Data

	General Motors			U.S. Steel			General Electric			Chrysler		
	Y	$X_{1,t-1}$	$X_{2,t-1}$	Y	$X_{1,t-1}$	$X_{2,t-1}$	Y	$X_{1,t-1}$	$X_{2,t-1}$	Y	$X_{1,t-1}$	$X_{2,t-1}$
1935	317.6	3078.5	2.8	209.9	1362.4	53.8	33.1	1170.6	97.8	40.29	417.5	10.5
36	391.8	4661.7	52.6	355.3	1807.1	50.5	45.0	2015.8	104.4	72.76	837.8	10.2
37	410.6	5387.1	156.9	469.9	2676.3	118.1	77.2	2803.3	118.0	66.26	883.9	34.7
38	257.7	2792.2	209.2	262.3	1801.9	260.2	44.6	2039.7	156.2	51.60	437.9	51.8
39	330.8	4313.2	203.4	230.4	1957.3	312.7	48.1	2256.2	172.6	52.41	679.7	64.3
40	461.2	4643.9	207.2	361.6	2202.9	254.2	74.4	2132.2	186.6	69.41	727.8	67.1
41	512.0	4551.2	255.2	472.8	2380.5	261.4	113.0	1834.1	220.9	68.35	643.6	75.2
42	448.0	3244.1	303.7	445.6	2168.6	298.7	91.9	1588.0	287.8	46.80	410.9	71.4
43	499.6	4053.7	264.1	361.6	1985.1	301.8	61.3	1749.4	319.9	47.40	588.4	67.1
44	547.5	4379.3	201.6	288.2	1813.9	279.1	56.8	1687.2	321.3	59.57	698.4	60.5
45	561.2	4840.9	265.0	258.7	1850.2	213.8	93.6	2007.7	319.6	88.78	846.4	54.6
46	688.1	4900.9	402.2	420.3	2067.7	232.6	159.9	2208.3	346.0	74.12	893.8	84.8
47	568.9	3526.5	761.5	420.5	1796.7	264.8	147.2	1656.7	456.4	62.68	579.0	96.8
48	529.2	3254.7	922.4	494.5	1625.8	306.9	146.3	1604.4	543.4	89.36	694.6	110.2
49	555.1	3700.2	1020.1	405.1	1667.0	351.1	98.3	1431.8	618.3	78.98	590.3	147.4
50	642.9	3755.6	1099.0	418.8	1677.4	357.8	93.5	1610.5	647.4	100.66	693.5	163.2
51	755.9	4833.0	1207.7	588.2	2289.5	342.1	135.2	1819.4	671.3	160.62	809.0	203.5
52	891.2	4924.9	1430.5	645.5	2159.4	444.2	157.3	2079.7	726.1	145.00	727.0	290.6
53	1304.4	6241.7	1777.3	641.0	2031.3	623.6	179.5	2371.6	800.3	174.93	1001.5	346.1
54	1486.7	5593.6	2226.3	459.3	2115.5	669.7	189.6	2759.9	888.9	172.49	703.2	414.9

TABLE 2.1 (Continued)

	Atlantic Richfield			I.B.M.			Union Oil		
	Y	$X_{1,t-1}$	$X_{2,t-1}$	Y	$X_{1,t-1}$	$X_{2,t-1}$	Y	$X_{1,t-1}$	$X_{2,t-1}$
1935	39.68	157.7	183.2	20.36	197.0	6.5	24.43	138.0	100.2
36	50.73	167.9	204.0	25.98	210.3	15.8	23.21	200.1	125.0
37	74.24	192.9	236.0	25.94	223.1	27.7	32.78	210.1	142.4
38	53.51	156.7	291.7	27.53	216.7	39.2	32.54	161.2	165.1
39	42.65	191.4	323.1	24.60	286.4	48.6	26.65	161.7	194.8
40	46.48	185.5	344.0	28.54	298.0	52.5	33.71	145.1	222.9
41	61.40	199.6	367.7	43.41	276.9	61.5	43.50	110.6	252.1
42	39.67	189.5	407.2	42.81	272.6	80.5	34.46	98.1	276.3
43	62.24	151.2	426.6	27.84	287.4	94.4	44.28	108.8	300.3
44	52.32	187.7	470.0	32.60	330.3	92.6	70.80	118.2	318.2
45	63.21	214.7	499.2	39.03	324.4	92.3	44.12	126.5	336.2
46	59.37	232.9	534.6	50.17	401.9	94.2	48.98	156.7	351.2
47	58.02	249.0	566.6	51.85	407.4	111.4	48.51	119.4	373.6
48	70.34	224.5	595.3	64.03	409.2	127.4	50.00	129.1	389.4
49	67.42	237.3	631.4	68.16	482.2	149.3	50.59	134.8	406.7
50	55.74	240.1	662.3	77.34	673.8	164.4	42.53	140.8	429.5
51	80.30	327.3	683.9	95.30	676.9	177.2	64.77	179.0	450.6
52	85.40	359.4	729.3	99.49	702.0	200.0	72.68	178.1	466.9
53	81.90	398.4	774.3	127.52	793.5	211.5	73.86	186.8	486.2
54	81.43	365.7	804.9	135.72	927.3	238.7	89.51	192.7	511.3

TABLE 2.1 (Continued)

	Westinghouse			Goodyear			Diamond Match		
	Y	$X_{1,t-1}$	$X_{2,t-1}$	Y	$X_{1,t-1}$	$X_{2,t-1}$	Y	$X_{1,t-1}$	$X_{2,t-1}$
1935	12.93	191.5	1.8	26.63	290.6	162	2.54	70.91	4.50
36	25.90	516.0	.8	23.39	291.1	174	2.00	87.94	4.71
37	35.05	729.0	7.4	30.65	335.0	183	2.19	82.20	4.57
38	22.89	560.4	18.1	20.89	246.0	198	1.99	58.72	4.56
39	18.84	519.9	23.5	28.78	356.2	208	2.03	80.54	4.38
40	28.57	628.5	26.5	26.93	289.8	223	1.81	86.47	4.21
41	48.51	537.1	36.2	32.08	268.2	234	2.14	77.68	4.12
42	43.34	561.2	60.8	32.21	213.3	248	1.86	62.16	3.83
43	37.02	617.2	84.4	35.69	348.2	274	.93	62.24	3.58
44	37.81	626.7	91.2	62.47	374.2	282	1.18	61.82	3.41
45	39.27	737.2	92.4	52.32	387.2	316	1.36	65.85	3.31
46	53.46	760.5	86.0	56.95	347.4	302	2.24	69.54	3.23
47	55.56	581.4	111.1	54.32	291.9	333	3.81	64.97	3.90
48	49.56	662.3	130.6	40.53	297.2	359	5.66	68.00	5.38
49	32.04	583.8	141.8	32.54	276.9	370	4.21	71.24	7.39
50	32.24	635.2	136.7	43.48	274.6	376	3.42	69.05	8.74
51	54.38	723.8	129.7	56.49	339.9	391	4.67	83.04	9.07
52	71.78	864.1	145.5	65.98	474.8	414	6.00	74.42	9.93
53	90.08	1193.5	174.8	66.11	496.0	443	6.53	63.51	11.68
54	68.60	1188.9	213.5	49.34	474.5	468	5.12	58.12	14.33

Y_{it} is gross investment for firm i in year t.

$X_{1i,t-1}$ is the value of firm i in year t - 1.

$X_{2i,t-1}$ is the stock of plant and equipment for firm i in year t - 1.

Gross investment equals additions to plant and equipment plus maintenance and repairs in millions of dollars. Value to the firm is the price of common and preferred shares at December 31 times the number of common and preferred shares outstanding plus total book value of debt. The stock of plant and equipment is the accumulated sum of net additions to plant and equipment minus depreciation allowance. All dollar values are deflated by an appropriate index using 1947 as base. For more details see Vinod and Ullah (1981, pp. 259-260) or Boot and deWit (1960).

In this section estimates of the production function parameters will be obtained using three of the methods discussed so far: separate equation OLS, classical pooling, and aggregation (Chapter 1). Later in the book these data will be used to illustrate other methods. Note that estimates using a particular method may not be appropriate for these data. The estimates are obtained purely for illustrative purposes.

The coefficients in equation (2.31) could be estimated using separate equation OLS applied to the time series for each firm. This can be achieved using any standard regression software package. Such routines are available in SPSS (1986), SAS (1982) and many other well known statistical software packages. In this example, MINITAB was used (see Ryan, Joiner, and Ryan, 1985). The estimates of each firm's regression coefficients are shown in Table 2.2 along with the t-statistics (in parentheses below each estimate), coefficient of determination for each regression (R^2) and the standard error of the regression (S).

Alternatively, suppose it was felt that the regression coefficients for each firm were equal, that is,

$$\begin{aligned} \beta_{01} &= \beta_{02} = \cdots = \beta_{0N} = \bar{\beta}_0 \\ \beta_{11} &= \beta_{12} = \cdots = \beta_{1N} = \bar{\beta}_1 \\ \beta_{21} &= \beta_{22} = \cdots = \beta_{2N} = \bar{\beta}_2 \end{aligned} \tag{2.32}$$

In this case the equation representing investment could be written

$$Y_{it} = \bar{\beta}_0 + \bar{\beta}_1 X_{1i,t-1} + \bar{\beta}_2 X_{2i,t-1} + \varepsilon_{it} \tag{2.33}$$

TABLE 2.2 Estimates of Investment Function Coefficients Using Separate Equation OLS, Classical Pooling, and Aggregation

	$\hat{\beta}_{0i}$	$\hat{\beta}_{1i}$	$\hat{\beta}_{2i}$	R^2(%)	S
General Motors	-149.8	0.119	0.371	92.1	91.8
	(- 1.42)	(4.62)	(10.02)		
U.S. Steel	- 49.7	0.171	0.409	48.1	95.5
	(- 0.34)	(2.32)	(2.83)		
General Electric	- 10.0	0.027	0.152	70.5	27.9
	(- 0.32)	(1.71)	(5.90)		
Chrysler	- 6.2	0.078	0.316	91.4	13.3
	(- 0.46)	(3.91)	(10.96)		
Atlantic Richfield	26.1	0.135	0.008	63.4	9.1
	(3.78)	(2.35)	(0.37)		
I.B.M.	- 8.7	0.131	0.085	95.2	8.1
	(- 1.91)	(4.22)	(0.85)		
Union Oil	- 4.5	0.088	0.124	76.4	9.4
	(- 0.40)	(1.34)	(7.26)		
Westinghouse	- 0.6	0.053	0.092	74.4	10.2
	(- 0.07)	(3.37)	(1.64)		
Goodyear	- 7.7	0.076	0.082	66.5	9.1
	(- 0.82)	(2.22)	(2.92)		
Diamond Match	0.2	0.004	0.437	64.1	1.1
	(0.09)	(0.16)	(5.47)		
CP	- 43.0	0.115	0.232	81.3	94.2
	(- 4.53)	(19.79)	(9.11)		
Aggregation	- 32.3	0.098	0.261	94.0	14.8
	(- 1.68)	(4.93)	(10.73)		

where $\bar{\beta}_0$, $\bar{\beta}_1$, and $\bar{\beta}_2$ represent the common regression coefficients. If equations (2.32) are true, the common coefficients can be estimated by classical pooling. The CP estimates of the coefficients are shown in the CP row of Table 2.2 along with t-statistics and the R-squared value and standard error of the pooled regression. The pooled regression is performed on all NT = 200 observations.

Finally, from Chapter 1, the aggregation method can be used to estimate the coefficients in equation (2.33) again assuming equations (2.32) are true.

These estimates were again obtained with the basic OLS regression routine in MINITAB. First the data are aggregated over all ten cross-sections as in equations (1.9) and (1.10). The aggregated data may be written as

$$\overline{Y}_t = \frac{1}{N} \sum_{i=1}^{N} Y_{it} \tag{2.34}$$

$$\overline{X}_{1,t-1} = \frac{1}{N} \sum_{i=1}^{N} X_{1i,t-1} \tag{2.35}$$

$$\overline{X}_{2,t-1} = \frac{1}{N} \sum_{i=1}^{N} X_{2i,t-1} \tag{2.36}$$

The regression equation to be estimated will then be

$$\overline{Y}_t = \overline{\beta}_0 + \overline{\beta}_1 \overline{X}_{1,t-1} + \overline{\beta}_2 \overline{X}_{2,t-1} + \overline{\varepsilon}_t \tag{2.37}$$

where there are a total of twenty observations on each variable. Applying OLS to this equation results in the estimates shown in the aggregation row of Table 2.2 again with t-statistics, the R-squared value and the standard error of the regression.

It is important to note that the results from CP and aggregation assume that all coefficients are equal. This is an assumption that can be tested as discussed in the previous section. The hypotheses may be stated as

$$H_0: \ \beta_1 = \beta_2 = \cdots = \beta_N = \overline{\beta}.$$

versus H_1: not all coefficient vectors are equal.

where $\beta_i = [\,\beta_{0i} \ \ \beta_{1i} \ \ \beta_{2i}]'$ is the vector of regression coefficients for firm i.

The test statistic is a special case of the F-statistic shown in equation (2.30):

$$F = \frac{(SSE_2 - SSE_1)/(NK - K)}{SSE_1/(NT - K)}$$

where SSE_1 is the residual sum of squares from the unrestricted model shown in equation (2.1) and SSE_2 is the residual sum of squares from the restricted model in equation (2.3). The null hypothesis shows the restrictions placed on the model. There are $NK - K = 27$ restrictions ($N = 10$ and $K = 3$). The other values necessary to compute the F-statistic are

$SSE_2 = 1{,}749{,}227$

$SSE_1 = 321{,}632$

$NT - K = 197$

The F-statistic value is 5.95. Under the null hypothesis (and the conditions specified in Section 2.2) the statistic will have a central F-distribution with 27 numerator and 197 denominator degrees of freedom. The observed value of the statistic has a p-value of 0.0000 so the null hypothesis would be rejected. It is inappropriate to assume all regression coefficients are equal. This result is not unexpected given the widely varying values of the separate equation OLS coefficient estimates observed in Table 2.2.

3
Seemingly Unrelated Regressions

3.1 THE SEEMINGLY UNRELATED REGRESSIONS (SUR) MODEL

Thus far, two approaches have been presented for working with pooled data. The first was to assume that coefficients differed between cross-sectional units and that the cross-sectional units were independent. This resulted in the analysis of each cross-section's regression equation separately. The second approach was to assume that coefficients were identical for each cross-sectional unit and to pool all NT observations into a single equation for estimation of the coefficients.

When the assumption of different coefficients between cross sections seems reasonable, it may still seem unreasonable to assume that each cross section behaves independently of the others. If it is assumed that disturbances measured at the same point in time may be correlated between different cross-sections, a more efficient estimator can be obtained than by simply applying least squares to each equation separately.

This correlation between disturbances of different cross-sections at the same point in time is called contemporaneous correlation. When several separate equation regressions are estimated, the disturbances at a given point in time might be expected to reflect some common unmeasurable or omitted influences. Contemporaneous correlation could be the result of these common factors which are not included in the regressions. Examples where such influences might occur are demand equations for different commodities, investment functions for firms or consumption functions for different groups. Marcis and Smith (1973), for example, estimated demand functions for cash and short-term Treasury obligations for firms in nine corporate asset size categories. Since each asset size group is subject to similar external effects (changing monetary policy, factor prices, etc.) it is likely that contemporaneous correlation exists. In this chapter, extensions of the classical pooling model which take account of contemporaneous correlation are considered.

Again write the model as in equation (2.1)

$$Y_i = X_i\beta_i + \varepsilon_i \tag{3.1}$$

for $i = 1,\ldots,N$ with the following assumptions

Assumption 3.1: $E(\varepsilon_i) = 0$.
Assumption 3.2: $E(\varepsilon_i\varepsilon_i') = \sigma_i^2 I_T$.
Assumption 3.3: $E(\varepsilon_i\varepsilon_j') = \sigma_{ij} I_T$.
Assumption 3.4: X_i is a fixed matrix in repeated samples.

Here it is assumed that the coefficient vectors, β_i, can differ for different individuals.

The equation (3.1) can be rewritten as

$$Y = X\beta + \varepsilon \tag{3.2}$$

where

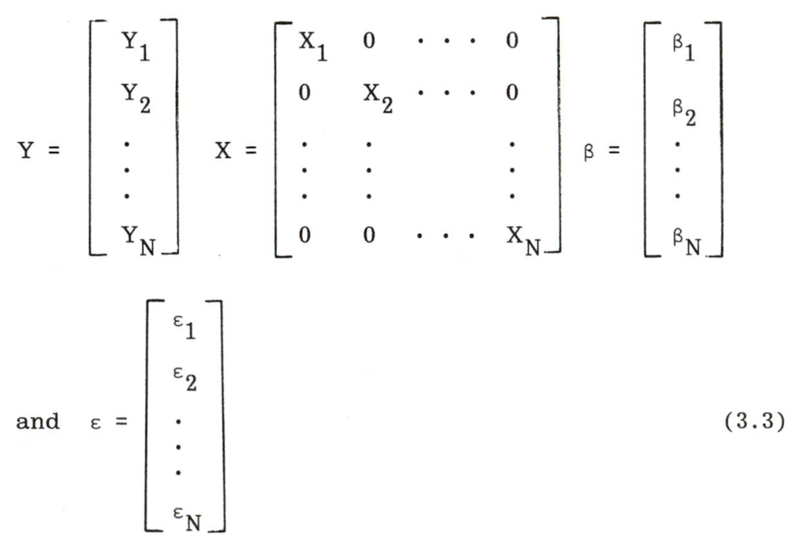

$$Y = \begin{bmatrix} Y_1 \\ Y_2 \\ \vdots \\ Y_N \end{bmatrix} \quad X = \begin{bmatrix} X_1 & 0 & \cdots & 0 \\ 0 & X_2 & \cdots & 0 \\ \vdots & \vdots & & \vdots \\ 0 & 0 & \cdots & X_N \end{bmatrix} \quad \beta = \begin{bmatrix} \beta_1 \\ \beta_2 \\ \vdots \\ \beta_N \end{bmatrix}$$

$$\text{and} \quad \varepsilon = \begin{bmatrix} \varepsilon_1 \\ \varepsilon_2 \\ \vdots \\ \varepsilon_N \end{bmatrix} \tag{3.3}$$

Assumptions (3.1) to (3.3) mean that the variance-covariance matrix of the disturbance vector ε in equation (3.2) can be written as

$$\Omega = \Sigma \otimes I \tag{3.4}$$

where

$$\Sigma = \begin{bmatrix} \sigma_1^2 & \sigma_{12} & \cdots & \sigma_{1N} \\ \sigma_{21} & \sigma_2^2 & \cdots & \sigma_{2N} \\ \vdots & \vdots & & \vdots \\ \sigma_{N1} & \sigma_{N2} & \cdots & \sigma_N^2 \end{bmatrix} \tag{3.5}$$

I is a $T \times T$ identity matrix, $\otimes$ denotes the Kronecker product, and σ_{ij} $(i \neq j)$ is the covariance of the disturbance terms from the ith and jth equation. The Kronecker product is defined as follows:

Suppose A is an $m \times n$ matrix with elements a_{ij} and B is a $p \times q$ matrix. Then the Kronecker product, $A \otimes B$, is

$$A \otimes B = \begin{bmatrix} a_{11}B & a_{12}B & \cdots & a_{1n}B \\ a_{21}B & a_{22}B & \cdots & a_{2n}B \\ \vdots & \vdots & & \vdots \\ a_{m1}B & a_{m2}B & \cdots & a_{mn}B \end{bmatrix}$$

If all the covariance terms, σ_{ij}, are zero, then OLS applied to each of the equations in (3.1) separately will provide an efficient estimator of each β_i. If contemporaneous correlation is present, however, it may be possible to construct more efficient estimators than the individual OLS estimators.

Generalized least squares estimation of the equation system in (3.2) produces the following estimator

$$\tilde{\beta}_{SUR} = (X'\Omega^{-1}X)^{-1} X'\Omega^{-1}Y \tag{3.6}$$

Dwivedi and Srivastava (1978) showed that whenever the X_i matrices span the same column space, OLS can be applied equation by equation without loss in efficiency. In practice this situation occurs when the independent variables for different individuals take exactly the same values as in Hirschey (1981). See also Gibbons (1982) and Binder (1983). In this case, the presence of contemporaneous correlation will not result in a loss of efficiency in the OLS estimators of the β_i.

Schmidt (1978) considers the case when a group of the same explanatory variables is excluded from certain of the equations. When SUR estimation is applied to the entire set of equations, the SUR estimates for the equations with the excluded variables will be the same as the OLS estimates for these equations. Revankar (1974) demonstrated this result in a two-equation case where the variables of one equation were a subset of the variables of the other.

When the X_i matrices do not span the same column space and contemporaneous correlation is present, the estimator $\tilde{\beta}_{SUR}$ will provide a more efficient estimator of the coefficient vector than will OLS applied separately to each equation.

Under assumptions (3.1) to (3.4), and adding the following assumption,

Assumption 3.5: $\lim_{T\to\infty} (X'\Omega^{-1}X)^{-1}$ is a finite nonsingular matrix,

the estimator $\tilde{\beta}_{SUR}$ is a best linear unbiased, consistent and asymptotically normally distributed estimator of β (see, e.g., Fomby et al., 1984, pp. 158). If ε in equation (3.2) is normally distributed, then $\tilde{\beta}_{SUR}$ is a minimum variance unbiased estimator and is asymptotically efficient.

When Ω is unknown, as will be the case in practice, an estimate of Ω must be computed and substituted into equation (3.6) to obtain the feasible Aitken estimator β.

The estimation of Ω (or, equivalent Σ) can proceed using residuals obtained either by (a) regressing the Y_i on the independent variables from the ith equation alone (the restricted residuals) or by (b) regressing each Y_i on all the independent variables in the system (3.2), (the unrestricted residuals). In either case the estimator $\hat{\Omega}$ can be used in (3.6) to construct a feasible Aitken estimator, provided $\hat{\Omega}^{-1}$ exists.

Writing the residuals (either restricted or unrestricted) as $\hat{\varepsilon}_{it}$, the disturbance variances and covariances can be estimated by

$$\hat{\sigma}_i^2 = \frac{1}{T - K} \sum_{t=1}^{T} \hat{\varepsilon}_{it}^2 \tag{3.7}$$

$$\hat{\sigma}_{ij} = \frac{1}{T - K} \sum_{t=1}^{T} \hat{\varepsilon}_{it} \hat{\varepsilon}_{jt} \tag{3.8}$$

If the number of explanatory variables differs between equations (K_i variables in each equation rather than K) equations (3.7) and (3.8) can be adjusted by using T in the denominator rather than T − K. If an unbiased estimator is desired there are possible degrees of freedom corrections but it is unclear whether they would provide an improvement in the estimators. See Judge et al. (1985), p. 469).

Substituting the estimators in equations (3.7) and (3.8) into Ω to obtain $\hat{\Omega}$ results in

$$\hat{\beta}_{SUR} = (X'\hat{\Omega}^{-1}X)^{-1} X'\hat{\Omega}^{-1} Y \tag{3.9}$$

The estimator, $\hat{\beta}_{SUR}$, was suggested by Zellner (1962) and will be referred to later as the ZEF estimator. An alternative estimator, also suggested by Zellner, is an iterative form of the ZEF estimator. This estimator will be called ITERZEF and is computed as follows: After obtaining $\hat{\beta}_{SUR}$ in equation (3.9), recompute the residuals and apply the formulas in (3.7) and (3.8) to obtain revised estimates of the σ_i^2 and σ_{ij}. Substitute these revised estimates into $\hat{\Omega}$ and recompute the coefficient estimates using (3.9). Continue this process until little change is observed in the successive estimates of the coefficients. Dhrymes (1971) has shown that the iterative estimator and the maximum likelihood estimator are equivalent. See also Oberhofer and Kmenta (1974) and Magnus (1978).

The feasible estimator $\hat{\beta}_{SUR}$ is consistent and asymptotically normally distributed under assumptions (3.1) to (3.5). If the disturbances are normally distributed, $\hat{\beta}_{SUR}$ will also be asymptotically efficient.

Parks (1967) considered the case when disturbances are both contemporaneously correlated and also follow a first-order autoregressive scheme over time. To assumptions (3.1) to (3.4) add

Assumption 3.6: $\varepsilon_{it} = \rho_i \varepsilon_{i,t-1} + u_{it}$

where

$$E(u_{it}) = 0$$

$$E(u_{it}^2) = \sigma_{ui}^2$$

$$E(\varepsilon_{i0}) = \frac{\sigma_{ui}^2}{1 - \rho_i^2}$$

The variance-covariance matrix for this model is

$$\Omega = \begin{bmatrix} \sigma_1^2 P_1 & \sigma_{12} P_2 & \cdots & \sigma_{1N} P_N \\ \sigma_{21} P_1 & \sigma_2^2 P_2 & \cdots & \sigma_{2N} P_N \\ \vdots & \vdots & & \vdots \\ \sigma_{N1} P_1 & \sigma_{N2} P_2 & \cdots & \sigma_N^2 P_N \end{bmatrix} \tag{3.10}$$

where

$$P_i = \begin{bmatrix} 1 & \rho_i & \rho_i^2 & \cdots & \rho_i^{T-1} \\ \rho_i & 1 & \rho_i & \cdots & \rho_i^{T-2} \\ \vdots & \vdots & \vdots & & \vdots \\ \rho_i^{T-1} & \rho_i^{T-2} & \rho_i^{T-3} & \cdots & 1 \end{bmatrix} \tag{3.11}$$

To estimate the ρ_i, each equation in (3.1) is estimated using OLS and the residuals, $\hat{\varepsilon}_{it}$, are used to estimate the ρ_i as in equation (2.12). The original data are transformed as in equations (2.14) and (2.15). Estimation now proceeds as previously described simply using the transformed data rather than the original data.

An alternative to this method, suggested by Kmenta and Gilbert (1970), is to first compute the ZEF parameter estimates and use the residuals based on these estimates in equation (2.12) to obtain estimates of the ρ_i. The original data can be transformed using equations (2.14) and (2.15). Estimation can then proceed as in the ZEF procedure. In subsequent sections, the two procedures discussed for estimating SUR equations with autocorrelated disturbances will be referred to as PARKS and KG, respectively. Of course, either of these methods could be iterated as in ITERZEF if desired.

Maeshiro (1980) suggested that the transformation to remove autocorrelation include the differentially transformed first observation as in equation (2.15). This will avoid any loss in efficiency due to omitting this observation. Doran and Griffiths (1983) used a Monte Carlo simulation to investigate the relative efficiencies of various estimators. Their results indicate that Maeshiro may have overstated the case for inclusion of the initial transformed observation. Although some loss in efficiency was observed when the observation was dropped, in most cases the efficiency loss was relatively small. Judge et al. (1985, pp. 485-490) suggest an alternative transformation for the first observation to the one in equation (2.15). They note that Park's transformation implicitly assumes that $E[u_{i1}u_{j1}] \neq \sigma_{ij}$. The Judge et al. transformation does not make this assumption. Given the Doran and Griffiths result it is unclear that much loss in efficiency will occur regardless of whether the initial observations are omitted from the analysis or transformed according to either the Parks or Judge et al. procedures.

Beach and MacKinnon (1979) studied ML estimation designed to take account of the initial observation and the stationarity condition on the error process. Their result applies only when the first-order autocorrelation coefficient is the same for all equations.

Guilkey and Schmidt (1973) studied SUR models with vector autoregressive errors while Hall (1982) examined models with vector autoregressive moving average (ARMA) errors. The Guilkey and Schmidt estimator is consistent and asymptotically efficient if no lagged values of the dependent variable appear as explanatory variables. If lagged variable values are included as explanatory variables the estimator is neither consistent nor asymptotically efficient. Spencer (1979) developed an estimator for this situation that possesses both of the optimal properties lacking in the Guilkey and Schmidt estimator. He compares his estimator to the PARKS estimator by a Monte Carlo simulation. Results suggest that PARKS will be superior if the coefficient of the lagged variable is large or if the autocorrelation is weak.

Otherwise, Spencer's estimator is superior. Wang, Hidiroglou, and Fuller (1980) developed an estimator similar to that of Spencer. The Wang et al. estimator is an adaptation of the two step Gauss-Newton method of estimating distributed lag models. See also Rausser and Oliver (1976).

As previously mentioned, when explanatory variables in each of the equations (3.1) are identical, LS estimation is efficient even when disturbances are contemporaneously correlated. However, when the number of time series observations differs between equations this result no longer holds. Conniffe (1985) examines this case and derives ML estimators of the regression coefficients under the assumption of normally distributed disturbances, derives variance estimates and discusses appropriate degrees of freedom for hypothesis tests. Swamy and Mehta (1975b) and Schmidt (1977) also study the SUR model with missing observations.

Rao (1974) and Srivastava and Srivastava (1983) consider SUR estimation when the model has been improperly specified, either by including an irrelevant variable or by excluding a relevant one.

Gallant (1975) proposed and estimator for nonlinear SUR equations.

Stanek and Koch (1985) demonstrated the equivalence of parameter estimates in growth curve models and the SUR model estimates based on unrestricted residuals. They point out the importance of the result for developing more general growth curve models.

Terasvirta (1975) considers prediction in SUR models. The vector to be predicted is denoted Y_p and the following assumptions are made:

$$Y_p = X_p \beta + \varepsilon_p \tag{3.12}$$

where X_p is known, $E(\varepsilon_p) = 0$, the variance-covariance matrix of ε_p is denoted by Ω_p, and the covariance between ε_p and the disturbance vector in the origianl set of equations is

$$\text{Cov}\ (\varepsilon, \varepsilon_p) = \Omega_{12} \tag{3.13}$$

Then the best linear unbiased predictor of Y_p will be

$$\hat{Y}_p = X_p \hat{\beta}_{SUR} + \Omega_{12}\Omega^{-1}(Y - X\hat{\beta}_{SUR}) \tag{3.14}$$

See also Baksalary and Kala (1979).

3.2 ANALYTIC RESULTS

SUR models and various estimators have been examined analytically in many articles. Zellner (1962) showed that his two-stage Aitken approach (ZEF) would be asymptotically more efficient than OLS applied equation by equation with maximum gain in efficiency occurring when contemporaneous correlation of the disturbances is high and explanatory variables in different equations are uncorrelated. For the normal case of two equations with X_2 a submatrix of X_1, Revankar (1974, 1976) derived exact finite-sample results that suggest that the gain in efficiency of ZEF over OLS may also occur in small samples $(T - K > 2)$ if the contemporaneous correlation is high. Revankar (1976) noted that the use of restricted (as opposed to unrestricted) residuals may lead to more efficient estimators of the β_i in small samples if the cross-equation correlations of disturbances are low. Zellner (1963), using restricted residuals and again using a two-equation system but with orthogonal regressors, $X_1'X_2 = 0$, obtained exact finite sample results suggesting that little loss in efficiency is observed for any degree of contemporaneous correlation or any sample size in using ZEF while considerable gains can be made. Definite gains are obtained for all sample sizes when $|\rho_c| > 0.3$, where ρ_c is the contemporaneous correlation coefficient for the disturbances in the two equations. Mehta and Swamy (1976) derived the exact second-order moments for a two-equation system using unrestricted residuals and no restrictions on the X_i matrices. Their results support using ZEF when $|\rho_c| \geq 0.3$ and the sample size, T, minus the number of distinct regressors, K, is greater than 23. When $|\rho_c| < 0.3$, the loss in efficiency is negligible for $T - K > 23$.

Kataoka (1974) derived the exact finite sample distribution of the estimator in a multiple equation system with orthogonal regressors and unrestricted residuals. Phillips (1977) and Kariya and Maekawa (1982) derived Edgeworth approximations to the exact distribution of the SUR estimator based on unrestricted residuals. Srivastava (1970) derived an approximation to the variance-covariance matrix of the ZEF estimator using restricted residuals which was later verified by Kakwani (1974). Srivastava and Upadhyaya (1978) obtained the same approximation for the ZEF estimator using unrestricted residuals. The two expressions were shown to be identical. See also Kariya (1981a).

Rothenberg (1984) developed approximations to the distribution of feasible GLS estimators when the disturbance covariance matrix depends on a small number of unknown parameters. He showed that the asymptotic normal distribution should serve as a reasonable approximation to the actual sampling distribution. This generalizes the result of Phillips (1977) which applied specifically to the SUR estimator.

Hillier and Satchell (1986) studied the finite sample properties of the estimator proposed by Telser (1964).

Phillips (1985) derived the exact finite sample distribution of the SUR estimator. His results generalized all presently known distribution theory for the SUR model.

Maekawa (1983), using nonorthogonal regressors and unrestricted residuals, expressed the distribution function of a two-equation system in terms of the canonical correlation coefficient between the regressors of both equations. The distribution of the estimator is shown to depend largely on the contemporaneous correlation of the disturbances rather than the canonical correlation between explanatory variables. Kunitomo (1977) examined the reduction in efficiency of the SUR estimator relative to OLS for different values of the correlation between explanatory variables. As the correlation increased, the relative efficiency decreased as suggested by Zellner (1962). See also Zellner and Huang (1962) and Binkley (1982).

Kakwani (1967) showed that Zellner's estimator is unbiased provided its expectation exists, the distribution of the disturbances is symmetrical, and the estimator of Ω is based on restricted residuals. Revankar (1974) showed that the result holds when the estimator of Ω is based on unrestricted residuals. Srivastava and Raj (1979) used results in Fuller and Battese (1973) to show that the expectation exists when the disturbance distribution is symmetrical with finite fourth-order moments and the expectation of $\hat{\Omega}^{-1}$ exists. See also Don and Magnus (1980).

Zellner (1971, pp. 240-246), Zellner and Vandaele (1975) and Srivastava (1973) investigated Bayesian approaches to estimation of the SUR model. Duncan (1983) developed an estimator that is more efficient than Zellner's SUR estimator when heteroskedasticity is present in at least one equation.

Srivastava and Srivastava (1984) propose an estimator that is a linear combination of the OLS estimator and the feasible GLS estimator. The OLS estimator receives less weight as the contemporaneous correlation increases. In a two-equation system with orthogonal regressors, the resulting estimator is asymptotically more efficient than either OLS or GLS. The optimal weights, however, involve the unknown contemporaneous correlation coefficients.

Tiao, Tan, and Chang (1977) studied a two-equation system with constraints on the coefficients.

3.3 SIMULATION RESULTS

Kmenta and Gilbert (1968) examined estimators of the SUR model parameters using a Monte Carlo simulation. They used both a two and four equation system with each equation having two explanatory variables. A variety of specifications of the disturbance term was used and the degree of correlation between the independent variables in separate equations was varied. The sample sizes used were T = 10, 20, and 100.

Four estimators were examined. These were ZEF, ITERZEF, an iterative estimator proposed by Telser (1964) and the ML estimator. The equations were also estimated by OLS to determine the extent to which efficiency gains were possible. Three of the estimators, ITERZEF, the Telser estimator and ML, provided numerically equivalent estimates. These three methods are simply different ways of obtaining the ML estimator as shown later by Dhrymes (1971). The comparison of alternative estimators, therefore, was simply a comparison of the ZEF versus ITERZEF estimators and the examination of the efficiencies of these estimators relative to OLS. See Oberhofer and Kmenta (1974) and Magnus (1978) for more on ML estimation for SUR models.

Little improvement was found in using the iterative procedure rather than the ZEF estimator. The ZEF estimator, however, performed better than OLS, in terms of relative efficiency, even in the small sample sizes used, when residuals were contemporaneously correlated. Of course, OLS was more efficient than ZEF when no contemporaneous correlation was introduced.

Several of the other asymptotic properties determined analytically were shown to hold in small samples. These include: 1. a small variance for the ZEF estimator as the degree of contemporaneous correlation increases, and 2. a smaller variance for ZEF as the cross-equation correlation of explanatory variables decreases.

The results of the experiment suggest that gains in efficiency will occur when contemporaneous correlation is taken into account by the estimation procedure. Also, the ZEF estimator appears to perform nearly as well as the iterative estimator and is considerably less complex to compute.

Kmenta and Gilbert (1970) performed a second simulation to determine the effects of first-order autocorrelation on estimators in SUR models. In this experiment a two-equation system with one explanatory variable in each equation was used. Sample sizes were 10, 20, and 100. First-order serial correlation was introduced in the generation of the disturbances.

Besides least squares, five other estimators were included in the study. These include the ZEF, PARKS, and KG estimators previously defined. In addition a procedure which used nonlinear least squares to estimate the autocorrelation coefficient, ρ, and then used the ZEF procedure on the transformed data, was examined. Also, a joint nonlinear estimation procedure was used which simultaneously estimated ρ and the other parameters of the model.

Results of the simulation suggest that the joint nonlinear procedure is superior to the other estimators when serial correlation is present. However, Kmenta and Gilbert point out that this procedure will be extremely expensive (in terms of computer time) unless the system of equations is a very small one. They suggest the KG procedure may be preferable due to its reduced computational complexity. Its performance was nearly as good as that of the joint nonlinear procedure.

As mentioned previously, Doran and Griffiths (1983) performed a simulation to determine the effect of omission or inclusion of the first observation when transforming to remove autocorrelation in a SUR system. Their results indicate the efficiency loss will typically be small if the first observation is omitted even in a sample size as small as 20.

3.4 INFERENCE IN THE SUR MODEL

The covariance matrix of the ZEF (or ITERZEF) estimator, $\hat{\beta}_{SUR}$, is given by

$$(X'\hat{\Omega}^{-1}X)^{-1} \tag{3.15}$$

The diagonal elements of the covariance matrix provide variances of the estimated regression coefficients. The standard errors are then available as the square roots of the diagonal elements and can be used in constructing t-statistics to test hypotheses about the population regression coefficients. That the tests statistics do have a t-distribution (with NT − NK degrees of freedom) depends on the asymptotic normality of $\hat{\beta}_{SUR}$. If the disturbances are normal or if the sample size is large enough, we can use the t-test to test hypotheses about the regression coefficients. How large is "large enough" is a matter for empirical investigation.

Srivastava and Tracy (1986) show that the standard formulae for estimating coefficient variances are biased when T is small and suggest alternative formulas.

Zellner (1962) also suggested a test of whether the coefficient vectors from the separate equations are equal. He called this a test for aggregation bias. The null hypothesis can be stated as

$$H_0: \ \beta_1 = \beta_2 = \cdots = \beta_N \tag{3.16}$$

If the null hypothesis is true, then when the data on individual units are aggregated into a single time series and estimation is performed as described in Section 1.4, the estimator will be unbiased. The test is not necessarily restricted to be a test of whether or not to aggregate the data, however, but also provides additional information on how individual units behave.

Assuming that the ε_i are normally distributed, Zellner derives the likelihood ratio test statistic to test the null hypothesis in (3.16). He shows that the LR test statistic is asymptotically equivalent to the following statistic

$$F = \frac{NT - NK}{(N-1)K} \times \frac{Y'\Omega^{-1}X(X'\Omega^{-1}X)^{-1}C'[C(X'\Omega^{-1}X)^{-1}C']^{-1}C(X'\Omega^{-1}X)^{-1}X'\Omega^{-1}Y}{Y'\Omega^{-1}Y - Y'\Omega^{-1}X(X'\Omega^{-1}X)^{-1}X'\Omega^{-1}Y} \tag{3.17}$$

where C is the matrix of the restrictions on the vector β; that is

$$C = \begin{bmatrix} I & -I & 0 & \cdots & 0 & 0 & 0 \\ 0 & I & -I & \cdots & 0 & 0 & 0 \\ \cdot & \cdot & \cdot & & \cdot & \cdot & \cdot \\ \cdot & \cdot & & & \cdot & \cdot & \cdot \\ \cdot & \cdot & \cdot & & \cdot & \cdot & \cdot \\ 0 & 0 & 0 & \cdots & I & -I & 0 \\ 0 & 0 & 0 & \cdots & 0 & I & -I \end{bmatrix} \tag{3.18}$$

The statistic F can be shown to have an F-distribution with $(N - 1)K$ numerator and $NT - NK$ denominator degrees of freedom. The statistic was also derived by Roy (1957, p. 82). Note that this statistic is simply the usual F-statistic for a set of general linear hypotheses set in the context of the SUR model. Zellner noted that $(N - 1)K \times F$ was distributed as χ^2 with $(N - 1)K$ degrees of freedom, asymptotically. Thus, either form of the test statistic could be used. Zellner suggests using the F-test in small samples. Of course in practice Ω must be replaced by a consistent estimator $\hat{\Omega}$. The statistic will then have the same distribution, asymptotically, as F computed with the true Ω matrix. Judge et al. (1985, p. 475) write the F-statistic in an alternative form. They also show the forms of the Wald, likelihood ratio, and Lagrange multiplier test statistics. Denote these, respectively, as λ_W, λ_{LR}, and λ_{LM}. Although λ_W, λ_{LR}, and λ_{LM} have, asymptotically, Chi-square distributions with $(N - 1)K$ degrees of freedom, their finite sample distributions will not be the same. As shown by Berndt and Savin (1977), the following relationship holds:

$$\lambda_W \geq \lambda_{LR} \geq \lambda_{LM}$$

so the rejection of H_0 in equation (3.16) will be more likely with λ_W while acceptance will be more likely with λ_{LM}. Which of these statistics performs best in small samples would be a topic for empirical investigation.

Breusch and Pagan (1980) developed the Lagrange multiplier (LM) test for contemporaneous correlation. The performance of the test was studied by Kariya (1981b). The hypothesis to be tested is

H_0: Ω is a diagonal matrix.

against the alternative that the off-diagonal elements of Ω are nonzero. If H_0 is accepted, OLS can be applied to each equation separately without a loss in efficiency. If H_0 is rejected, then a procedure which takes account of the contemporaneous correlation should provide relatively more efficient estimators of the regression coefficients.

To construct the LM test statistic Breusch and Pagan suggest the following procedure:

Estimate each of the N equations in (3.1) by OLS and obtain OLS residuals

$$\hat{\varepsilon}_i = Y_i - X_i\hat{\beta}_i \tag{3.19}$$

where $\hat{\beta}_i$ is the OLS estimator of β_i. Estimate the σ_i^2 by

$$\hat{\sigma}_i^2 = \frac{\hat{\varepsilon}_i'\hat{\varepsilon}_i}{T - K} \tag{3.20}$$

where the $\hat{\varepsilon}_i$ are residuals from the OLS regressions, and construct the LM statistic as

$$LM_{SUR} = T \sum_{i=1}^{N} \sum_{j=1}^{i-1} r_{ij}^2 \qquad (3.21)$$

where

$$r_{ij} = T^{-1}(\hat{\sigma}_i^2 \hat{\sigma}_j^2 \hat{\varepsilon}_i \hat{\varepsilon}_j)$$

The statistic LM_{SUR} will be distributed as $\chi^2(\frac{1}{2}N\ (N - 1))$ if the null hypothesis is true.

As an alternative to the LM test, the likelihood ratio test can be constructed as

$$LR_{SUR} = \frac{|\Omega_0^*|}{|\Omega^*|} \qquad (3.22)$$

where Ω_0 is the variance-covariance matrix under the null hypothesis, Ω is the unrestricted variance-covariance matrix, the *'s indicate maximum likelihood estimators and the vertical lines indicate determinants. Under the null hypothesis, LR_{SUR}, will have the same asymptotic distribution as LM_{SUR}.

Since these two statistics have the same distribution asymptotically the choice between the two must be made on the basis of superior small sample performance or computational simplicity. The LM_{SUR} statistic appears to be simplest to compute. Evidence on how the two statistics compare in small samples is limited. Harvey and Phillips (1982) provide some evidence that the LM_{SUR} statistic will have an empirical level of significance closer to the nominal level than will the LR_{SUR} statistic in small samples. Their results, however, are obtained in the special case of testing whether one equation in a system such as that of (3.1) should be considered part of the system for purposes of estimation.

Conniffe (1982a) provides a test of whether the explanatory variables are independent of the disturbances. Rao (1974) studies the case when an important explanatory variable has been omitted from the regression equations. If such an omitted variable is correlated with the included explanatory variables then the situation investigated by Conniffe may occur. Rao has shown that in the case of such misspecification, SUR would be more in error, in terms of both bias and variance, than would be least squares.

McElroy (1977a) developed a goodness of fit measure for the SUR model computed as

$$R_{SUR}^{2} = 1 - \frac{\hat{\varepsilon}'\left(\hat{\sum}^{-1} \otimes I\right)\hat{\varepsilon}}{y'\left(\hat{\sum}^{-1} \otimes I\right)y} \tag{3.23}$$

where $\hat{\varepsilon}$ is the vector of residuals from the estimated SUR model. The R_{SUR}^2 statistic is related to the F-statistic used for testing whether all regression coefficients in the system are zero and its values do lie between zero and one. See also Buse (1979).

To test whether first-order autocorrelation is present,

$$H_0: \rho_1 = \rho_2 = \cdots = \rho_N = 0 \tag{3.24}$$

Harvey suggested the following modified Lagrange multiplier statistic

$$T \sum_{j=1}^{N} \hat{\rho}_j^{\,2} \tag{3.25}$$

where the $\hat{\rho}_j$ are estimates of the ρ_j computed as in equation (2.12). Under the null hypothesis in equation (3.24) the test statistic has a chi-square distribution with N degrees of freedom.

Guilkey (1974) developed two test statistics to test for first-order vector autoregressive disturbances which could be used to test the less general hypothesis in equation (3.24) as well.

See also Section 9.2 of this book.

3.5 APPLICATIONS AND SOFTWARE

Beckwith (1972) and Wildt (1974) used the SUR model to examine the sales response of competing brands to advertising expenditures. The ZEF and ITERZEF procedures were compared.

Marcis and Smith (1973) studied the demand for liquid asset balances by U.S. manufacturing corporations. Marcis and Smith (1974) also use the SUR model to analyze the determinants of movements in short-term interest rates for seven countries. Albon and Valentine (1977) studied demand for bank loans in Australia.

Carlson (1978) used SUR to estimate demand for automobiles of different sizes. Cross-sectional units in this study are the size of the car, that is, subcompact, compact, intermediate, full size, and luxury. Carlson and Umble (1980) provide an update of the study. In both of these articles, forecasts from the SUR estimated models are considered.

Jalilvand and Harris (1984) examine the issuance of long-term debt, short-term debt and equity, the maintenance of corporate liquidity, and the payment of dividends. The speed of adjustment of the firms to long-run financial targets is examined. Barth, Kraft, and Kraft (1979) used the SUR approach to estimate a price equation for the manufacturing sector of the United States. The cross-sectional units are manufacturing industries. A study to determine the price elasticities of electricity was conducted by Wilson (1974). Griffin (1977) used the SUR model to estimate interfuel substitution relationships between fossil fuels in the generation of electricity.

Brown and Kadiyala (1985) suggest applying the SUR model to event studies in finance and economics. Their work involves joint estimation of the SUR parameters and certain missing dependent variable observations as studied in Brown and Kadiyala (1983).

Related papers include Leonard (1982), Nelson and Wohar (1983), Brehm and Saving (1964), Lin (1985) and Lee (1976).

Other surveys of the seemingly unrelated regression model can be found in Judge et al. (1985, pp. 465-514) and Srivastava and Dwivedi (1979). An excellent and extensive survey of the theoretical results associated with the SUR model is Srivastava and Giles (1987).

White (1978) developed an econometric package called SHAZAM which includes a routine for SUR estimation. The RATS (Regression Analysis of Time Series) econometric software package also includes a routine for SUR estimation. RATS is distributed through TSP International. A program for estimating Gallant's (1975) nonlinear SUR model was reported on by Faurot and Fon (1978). The latest version of SAS contains routines for estimating both linear and nonlinear SUR.

3.6 EXAMPLE: INVESTMENT FUNCTION ESTIMATION

The data discussed in Section 2.3 will be used again in this section to illustrate the SUR procedure. The equations to be estimated can be written

$$Y_{it} = \beta_{0i} + \beta_{1i}X_{1i,t-1} + \beta_{2i}X_{2i,t-1} + \varepsilon_{it} \tag{3.26}$$

where Y_{it}, $X_{1i,t-1}$ and $X_{2i,t-1}$ are defined in Section 2.3. The coefficients β_{0i}, β_{1i}, and β_{2i} are initially assumed to be different for each cross-sectional unit (although this assumption can be tested).

For illustrative purposes, only the first two firms, General Motors and U.S. Steel will be used. If no contemporaneous correlation exists between the disturbances for these two firms, OLS is the appropriate estimation technique. If disturbances are contemporaneously correlated, SUR estimation should provide more efficient estimates of the regression coefficients.

Table 3.1 shows the results of estimating equations (3.26) using OLS and the ZEF procedure described earlier in the chapter. Restricted residuals were used in computing the variance-covariance matrix. The variance covariance matrix can be written as in equation (3.4). For this example, the estimate of Σ is

TABLE 3.1 Estimates of Investment Function Coefficients Using Separate Equation OLS and ZEF

	$\hat{\beta}_{0i}$ (Standard error) (t-stat)	$\hat{\beta}_{1i}$ (Standard error) (t-stat)	$\hat{\beta}_{2i}$ (Standard error) (t-stat)
General Motors			
OLS	-149.8 (105.8) (-1.42)	0.119 (0.0258) (4.62)	0.371 (0.0371) (10.02)
ZEF	-148.9 (103.1) (-1.45)	0.117 (0.0251) (4.67)	0.384 (0.0366) (10.51)
U.S. Steel			
OLS	-49.7 (146.6) (-0.34)	0.171 (0.0736) (2.32)	0.409 (0.1449) (2.83)
ZEF	-19.5 (142.7) (-0.14)	0.153 (0.0716) (2.14)	0.428 (0.1434) (2.99)

$$\hat{\Sigma} = \begin{bmatrix} \hat{\sigma}_1^2 & \hat{\sigma}_{12} \\ \hat{\sigma}_{21} & \hat{\sigma}_2^2 \end{bmatrix} = \begin{bmatrix} 8423.88 & -2617.40 \\ -2617.40 & 9116.40 \end{bmatrix} \quad (3.27)$$

See equations (3.7) and (3.8) for computation of the variances and covariances. The contemporaneous correlation can be estimated by

$$\begin{aligned} \hat{\rho}_c &= \frac{\hat{\sigma}_{12}}{(\hat{\sigma}_1^2 \hat{\sigma}_2^2)^{\frac{1}{2}}} \\ &= \frac{-2617.40}{[(8423.88)(9116.10)]^{\frac{1}{2}}} \quad (3.28) \\ &= -0.30 \end{aligned}$$

In Table 3.1 the coefficient estimates for both the OLS and ZEF procedures are shown. Below these estimates in parentheses are the standard errors using each technique and the t-statistics for testing the significance of the individual coefficients. Note that the standard errors from ZEF are slightly smaller than those from OLS, resulting from the gain in efficiency of ZEF over OLS. The gain in this example is fairly small, however. This may be due to the combination of a relatively low contemporaneous correlation of the disturbances ($\hat{\rho} = -0.3$) and a relatively high correlation between the independent variables.

4
Error Components Models

4.1 THE ANALYSIS OF COVARIANCE MODEL

In an attempt to relax the stringent assumption that individual coefficients remain constant and yet to improve on the technique of separate equation regressions, several models have been proposed. The models discussed in this chapter allow the equation intercepts to vary as a way of representing individual or time effects. These individual or time effects are typically thought to arise from the omission of important variables whose explicit inclusion in the model was not possible. These variables are often termed unmeasurable or unobservable effects. For example, Griliches (1977) and Hausman and Taylor (1981), among others, studied the benefits of education. Using wages as dependent variable and schooling as an independent variable, they noted that ability or ambition was an unmeasurable variable which varied across individuals but was relatively constant over time. Exclusion of this unobservable variable could produce biases in the estimates of other coefficients.

Chang (1978), in a study of the location of cotton textile mills, noted that the factors affecting industry location are numerous and complex and some would likely not be readily observable or measurable. He allowed variation in the equation intercepts to capture the individual and time effects.

The first of the models examined in this chapter, called the <u>analysis of covariance</u> (ANCOVA) or <u>least squares with dummy</u>

variables (LSDV) model, attempts to improve the specification of the classical pooling model by the introduction of dummy variables. The use of the dummy variables is an attempt to adjust for the missing independent variables which produce the individual and time effects. The equality of the slope coefficients from one cross section to another is accepted, but it is assumed that the intercepts differ.

Kuh (1959) was one of the first to propose a structure such as the one discussed in this chapter for treating pooled cross-sectional and time series data. Kuh pointed out that profound differences often existed between regression parameter estimates computed using data aggregated over time versus data aggregated over individuals. He suggested that the differences observed in these estimates may be due to the fact that cross-sectional data typically reflect long-run behavior while time series data will reflect adjustments in the shorter run. Kuh suggested that, to understand the behavior of individuals over time, a number of time series observations on each individual should be obtained. He also proposed an error structure similar to the one discussed in this chapter under the analysis of covariance and error components models headings.

Researchers who were among the first to use the analysis of covariance approach for analyzing pooled data were Mundlak (1961, 1963), Hoch (1962), and Johnson (1960, 1964).

The CP model of equation (2.3) can be rewritten as

$$Y = b_0 \iota_{NT} + Wd + \varepsilon \tag{4.1}$$

where the constant term, b_0, has been separated out, ι_{NT} represents an NT × 1 vector of ones, W is the matrix Z in equation (2.3) with the column of ones removed, that is

$$Z = [\iota_{NT} \,\vdots\, W] \tag{4.2}$$

and d is the vector of slope coefficients.

The elements of ε_{it} in the disturbance vector ε are often written

$$\varepsilon_{it} = c_i + s_t + v_{it} \tag{4.3}$$

The c_i represent time-invariant individual effects, the s_t are time period effects, and the v_{it} are remaining random effects. As discussed earlier, these effects can be thought of as arising from the omission from equation (4.1) of certain important variables which are unmeasurable or unobservable. The c_i represent the net effect of omitted time-invariant individual variables, for example, ability or personal background variables. The s_t represent the net effect of individual-invariant time effects such as prices or interest rates. The v_{it}, as always, represent the net effect of omitted variables which vary over both individuals and time.

Consider the following assumptions about the terms in equation (4.3):

Assumption 4.1: The c_i's are fixed parameters with $\sum_{i=1}^{N} c_i = 0$.

Assumption 4.2: The s_t's are fixed parameters with $\sum_{t=1}^{T} s_t = 0$.

Assumption 4.3: The v_{it} are independent and identically distributed random disturbances with

$$E(v_{it}) = 0,$$

$$E(v_{it}^2) = \sigma_v^2$$

$$E(v_{it}\, v_{jt}) = E(v_{is}\, v_{jt}) = E(v_{it}\, v_{is}) = 0$$

for all $i \neq j$, $s \neq t$.

In addition assume that W is fixed in repeated samples on Y and that W is of full rank.

These assumptions are commonly found in analyses using the ANCOVA or LSDV procedure.

The ANCOVA model can be estimated by including dummy variables in equation (4.1) to represent the time and individual effects and applying OLS to the equation. However, computationally more efficient methods of estimation are available. The procedure commonly used is to "sweep out" the constants c_i and s_t by applying the covariance transformation discussed in Swamy (1971), pp. 64-66). This procedure can be implemented as follows:

1. Find the mean of the time series observations for each individual separately.
2. Find the mean of the cross-sectional observations separately for each time period.

3. Find the overall mean of all the observations.
4. Transform the vector Y and the matrix W in (4.1) by subtracting out the appropriate time series and cross-sectional means (from steps 1 and 2) and then add in the overall mean (from step 3) to all observations.
5. Denote the vector of transformed dependent variable values as Y_1 and the matrix of transformed independent variable values as W_1. Now apply OLS to the transformed data. Call the estimator of the slope coefficients obtained in this manner $\hat{d}_{COV}$:

$$\hat{d}_{COV} = (W_1'W_1)^{-1} W_1'Y_1 \tag{4.4}$$

This is the ANCOVA estimator of the slope parameters. In matrix notation let

$$Q = I_{NT} - I_N \otimes \frac{\iota_T \iota_T'}{T} - \frac{\iota_N \iota_N'}{N} \otimes \iota_T + \frac{\iota_{NT} \iota_{NT}'}{NT} \tag{4.5}$$

where ι denotes a vector of ones of the subscripted size and $\otimes$ is the Kronecker product defined in Chapter 3. Then the ANCOVA estimator can be written as

$$\hat{d}_{COV} = (W'Q\ W)^{-1}\ W'Q\ Y \tag{4.6}$$

This is equivalent to the estimator defined in equation (4.4). Here Q is the "sweep" operator or the covariance transformation matrix. The estimator $\hat{d}_{COV}$ in equation (4.6) can be obtained by applying least squares to the equation

$$QY = QWd + Q\varepsilon \tag{4.7}$$

where $QY = Y_1$ and $QW = W_1$ as previously defined.

Note that one does not have to have both individual and time-effect parameters in the model. The more general case is presented here with models including only individual effects or only time effects viewed as special cases of the above model.

See, for example, Benus, Kmenta and Shapiro (1976), Johnson and Oksanen (1974), Palda and Blair (1970), or Ashenfelter (1978).

4.2 THE ERROR COMPONENTS MODEL WITH TIME AND INDIVIDUAL COMPONENTS

The problems existing with the covariance model are clearly evident. First, the slope coefficients are still assumed constant for all firms. Second, there are a substantial number of parameters to be estimated, thus using up a large number of degrees of freedom. And, finally, the coefficients of the dummy variables are not easily interpretable. They represent, as Maddala (1971b) pointed out, "some ignorance—just like the residuals v_{it}." Variables that might cause the regression line to shift are not specified; dummy variables are simply inserted to measure such shifts. Maddala then suggested that this "specific ignorance" might just as well be treated in a manner similar to that of our "general ignorance" (v_{it}). In other words, the coefficients c_i and s_t could be viewed as normally distributed random variables with mean zero and unknown variance. As well as achieving consistency in the incorporation of our "ignorance" into the model, this approach would also decrease the number of parameters to be estimated. Such a model has become known as the variance components or error components model.

Using the model as stated in equations (4.1) and (4.3), Assumptions 4.1 and 4.2 would be changed as follows:

Assumption 4.4: The c_i's in (4.3) are independent and identically distributed with mean zero and variance σ_c^2.

Assumption 4.5: The s_t's in (4.3) are independent and identically distributed with mean zero and variance σ_s^2.

Maintain Assumption 4.3 as previously stated but add the additional assumption:

Assumption 4.6: The c_i's, s_t's and v_{it}'s are mutually independent.

These are assumptions typically used when specifying a model with both time (s_t) and individual (c_i) error components.

Note that Kuh (1959) suggested a decomposition of the disturbances as

$$\varepsilon_{it} = c_i + v_{it} \tag{4.8}$$

but assumed that the components c_i and v_{it} may be correlated. Under this assumption Nerlove (1967) has shown that the same form of the covariance matrix results as with Assumptions (4.4) to (4.6).

The variance-covariance matrix can be written as

$$E(\varepsilon\varepsilon') = \Omega = \begin{bmatrix} \sigma_c^2 C_T & \sigma_s^2 I_T & \cdots & \sigma_s^2 IT \\ \sigma_s^2 I_T & \sigma_c^2 C_T & \cdots & \sigma_s^2 I_T \\ \vdots & \vdots & & \vdots \\ \sigma_s^2 I_T & \sigma_s^2 I_T & \cdots & \sigma_c^2 C_T \end{bmatrix} \quad (4.9)$$

where

$$C_T = \begin{bmatrix} \frac{\sigma^2}{\sigma_c^2} & 1 & \cdots & 1 \\ 1 & \frac{\sigma^2}{\sigma_c^2} & \cdots & 1 \\ \vdots & \vdots & & \vdots \\ 1 & 1 & \cdots & \frac{\sigma^2}{\sigma_c^2} \end{bmatrix} \quad (4.10)$$

I_T is a $T \times T$ identity matrix and $\sigma^2 = \sigma_c^2 + \sigma_s^2 + \sigma_v^2$.

The Ω matrix can also be written as

$$\Omega = \sigma_v^2 I_{NT} + \sigma_c^2 A + \sigma_s^2 B \quad (4.11)$$

where I_{NT} is an $NT \times NT$ identity matrix, A is an $NT \times NT$ matrix defined as

$$A = \begin{bmatrix} J_T & 0 & \cdots & 0 \\ 0 & J_T & \cdots & 0 \\ \vdots & \vdots & & \vdots \\ 0 & 0 & \cdots & J_T \end{bmatrix} \quad (4.12)$$

J_T is a T × T matrix of ones,

$$B = \begin{bmatrix} I_T & I_T & \cdots & I_T \\ I_T & I_T & \cdots & I_T \\ \cdot & \cdot & & \cdot \\ \cdot & \cdot & & \cdot \\ \cdot & \cdot & & \cdot \\ I_T & I_T & \cdots & I_T \end{bmatrix} \tag{4.13}$$

and I_T is a T × T identity matrix.

The GLS estimator of the parameters in (4.1) is

$$\tilde{d}_{EC} = (Z'\Omega^{-1}Z)^{-1}\, Z'\Omega^{-1}Y \tag{4.14}$$

where Z is as shown in equation (4.2).

A convenient form for the Ω^{-1} matrix is derived from equation (4.11) as

$$\Omega^{-1} = \frac{1}{\sigma_V^2}(I_{NT} - \gamma_1 A - \gamma_2 B + \gamma_3 J_{NT}) \tag{4.15}$$

where I_{NT}, A, and B are as previously defined, J_{NT} is an NT × NT matrix of ones, and the γ_i are functions of the variance components:

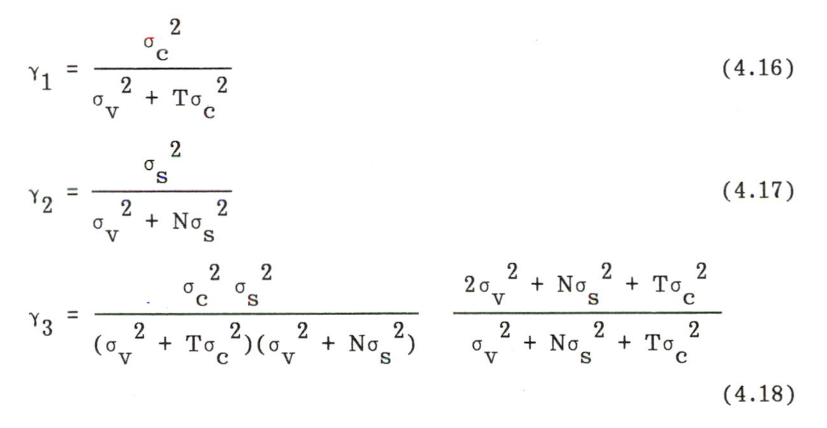

$$\gamma_1 = \frac{\sigma_C^2}{\sigma_V^2 + T\sigma_C^2} \tag{4.16}$$

$$\gamma_2 = \frac{\sigma_S^2}{\sigma_V^2 + N\sigma_S^2} \tag{4.17}$$

$$\gamma_3 = \frac{\sigma_C^2\, \sigma_S^2}{(\sigma_V^2 + T\sigma_C^2)(\sigma_V^2 + N\sigma_S^2)} \quad \frac{2\sigma_V^2 + N\sigma_S^2 + T\sigma_C^2}{\sigma_V^2 + N\sigma_S^2 + T\sigma_C^2} \tag{4.18}$$

Swamy (1971, p. 70) wrote the GLS estimator of d in equation (4.14) in the following form

$$\tilde{d}_{EC} = \left(\frac{W_1'W_1'}{\sigma_v^2} + \frac{W_2'W_2}{\sigma_2^2} + \frac{W_3'N_3W_3}{\sigma_3^2} \right)^{-1} \times \left(\frac{W_1'Y_1}{\sigma_v^2} + \frac{W_2'Y_2}{\sigma_2^2} + \frac{W_3'N_3Y_3}{\sigma_3^2} \right) \tag{4.19}$$

where

$$\sigma_3^2 = T\sigma_c^2 + \sigma_v^2$$

$$\sigma_2^2 = N\sigma_s^2 + \sigma_v^2$$

$$N_3 = I_N - \frac{\iota_N \iota_N'}{N}$$

W_1, W_2, and W_3 are matrices of transformed observations of the independent variables, Y_1, Y_2, and Y_3 are vectors of transformed observations of the dependent variables, and ι_N is an $N \times 1$ vector of ones. This form of the estimator for the d vector is particularly convenient to use in highlighting the computational procedure for the error components model and for discussing the differences that exist between the large number of possible estimation procedures that have been suggested for the model.

The matrices W_1 and Y_1 are as defined in Section 4.1. These are the transformed observations used in obtaining the analysis of covariance estimates of d.

To obtain W_2 and Y_2, find the mean of the time series observations separately for each unit and subtract out the time series mean from each observation for the respective unit. Average these deviations over units, separately, for each time period. To obtain W_3 and Y_3, average the observations over time, separately, for each unit. See also Nerlove (1971a).

The GLS estimator in (4.19) is the optimal matrix weighted average of the OLS regression estimators obtained by regressing Y_1 on W_1, Y_2 on W_2, and N_3Y_3 on N_3W_3. The estimator obtained from regressing Y_1 on W_1 is referred to as the "within" estimator since it utilizes variation within each cross-sectional unit or individual. The estimator obtained from regressing Y_2 on W_2 is referred to as the "between time period" estimator since it utilizes only variation between time periods. The estimator obtained from regressing N_3Y_3 on N_3W_3 is referred to as the "between group" estimator since it utilizes only variation between cross-sectional units. When the time effects, s_t, are omitted from the model (as discussed in the subsequent section, 4.3) the latter estimator is often simply called the between estimator.

The operational form of the error components estimator depends on how estimators for σ_2^2, σ_3^2, and σ_v^2 (or, equivalently, σ_c^2, σ_s^2, and σ_v^2) are constructed. Several estimators of the variance components have been proposed in the literature.

Wallace and Hussain (1969) used estimators of the variance components based on observed residuals obtained by applying OLS directly to (4.1). The OLS residuals, $\hat{\varepsilon}_i$, are then used to estimate σ_c^2, σ_s^2, and σ_v^2 as follows:

$$\hat{\sigma}_v^2 = \frac{1}{(N-1)(T-1)} \sum_{i=1}^{N} \sum_{t=1}^{T} \left(\hat{\varepsilon}_{it} - \frac{1}{T} \sum_{t=1}^{T} \hat{\varepsilon}_{it} - \frac{1}{N} \sum_{i=1}^{N} \hat{\varepsilon}_{it} \right)^2 \tag{4.20}$$

$$\hat{\sigma}_c^2 = \frac{1}{T} \left[\frac{1}{(N-1)T} \sum_{i=1}^{N} \left(\sum_{t=1}^{T} \hat{\varepsilon}_{it} \right)^2 - \hat{\sigma}_v^2 \right] \tag{4.21}$$

$$\hat{\sigma}_s^2 = \frac{1}{N} \left[\frac{1}{N(T-1)} \sum_{t=1}^{T} \left(\sum_{i=1}^{N} \hat{\varepsilon}_{it} \right)^2 - \hat{\sigma}_v^2 \right] \tag{4.22}$$

Wallace and Hussain showed these estimators to be consistent under the assumption that the independent variables are nonstochastic. However, Amemiya (1971) showed the estimators were not consistent if a lagged value of the dependent variable was used as an explanatory variable. Henderson (1971) also states that the estimators may be biased since OLS residuals are used. He suggests that the use of unbiased estimators may have important implications in small samples.

Amemiya (1971) suggests using residuals from the ANCOVA estimation of (4.1) in place of the OLS residuals used by Wallace and Hussain. Otherwise, Amemiya's variance estimators would be computed as in equations (4.20), (4.21), and (4.22). Amemiya shows that, asymptotically, the Wallace and Hussain variance estimators will have a larger covariance matrix than that of the variance estimator based on ANCOVA residuals.

Amemiya derives maximum likelihood estimators (MLE) of the variance components under the assumption that the disturbances are normally distributed. He shows that the MLE's and the estimators from (4.20), (4.21), and (4.22) with OLS residuals replaced by ANCOVA residuals are asymptotically equivalent. He also shows that the MLE's are consistent and asymptotically normal.

Swamy and Arora (1972) develop variance estimators using the mean squared residuals of the within groups, between time periods and between groups estimators. These estimators can be written as

$$\hat{\sigma}_v^2 = \frac{Y_1'M_1Y_1}{(N-1)(T-1)-K+1}$$

$$= \sum_{i=1}^{N}\sum_{t=1}^{T} \frac{\hat{\varepsilon}_{1it}^2}{(N-1)(T-1)-K+1} \tag{4.23}$$

where $M_1 = I_{(N-1)(T-1)} - W_1(W_1'W_1)^{-1}W_1'$. Thus σ_v^2 is estimated by using the <u>mean squared error</u> (MSE) from the regression of Y_1 on W_1. The $\hat{\varepsilon}_{1it}$ are residuals from this regression;

$$\hat{\sigma}_2^2 = \frac{Y_2'M_2Y_2}{T-K}$$

$$= \sum_{t=1}^{T} \frac{\hat{\varepsilon}_{2t}^2}{T-K} \tag{4.24}$$

where $M_2 = I_T - W_2(W_2'W_2)^{-1}W_2'$. Thus σ_2^2 is estimated by using the MSE from the regression of Y_2 on W_2. The $\hat{\varepsilon}_{2t}$ are residuals from this regression;

$$\hat{\sigma}_3^2 = \frac{Y_3'M_3Y_3}{N-K}$$

$$= \sum_{i=1}^{N} \frac{\hat{\varepsilon}_{3i}^2}{N-K} \tag{4.25}$$

where $M_3 = I_N - W_3(W_3'W_3)^{-1}W_3'$. Thus σ_3^2 is estimated by using the MSE from the regression of Y_3 on W_3.

Rao (1972) developed minimum norm quadratic unbiased estimators (MINQUE) that can be used to estimate the variance components. The MINQUE method was developed by Rao (1970) to estimate heteroskedastic variances in linear models and extended in subsequent articles (Rao 1971a, 1971b for example).

Nerlove (1971b) uses the following variance component estimators:

$$\hat{\sigma}_v^2 = \sum_{i=1}^{N} \sum_{t=1}^{T} \frac{\hat{\varepsilon}_{it}^2}{NT} \tag{4.26}$$

where the $\hat{\varepsilon}_{it}$ are obtained from the ANCOVA estimation.

$$\hat{\sigma}_c^2 = \sum_{i=1}^{N} \frac{(\hat{c}_i - \hat{\bar{c}})^2}{N} \tag{4.27}$$

where the $\hat{c}_i$ are the estimated coefficients of the dummy variables for individuals obtained from the ANCOVA estimation and $\hat{\bar{c}}$ is the average of those coefficients.

Similarly,

$$\hat{\sigma}_s^2 = \sum_{t=1}^{T} \frac{(\hat{s}_t - \hat{\bar{s}})^2}{T} \tag{4.28}$$

where the $\hat{s}_t$ are the estimated coefficients of the time dummy variables.

Fuller and Battese (1974) used a Henderson Method III type estimator of the variance components. Their model applies under somewhat more general assumptions than those presented earlier but their estimators under the assumptions presented here are:

$$\hat{\sigma}_v^2 = \sum_{i=1}^{N} \sum_{t=1}^{T} \frac{\hat{\varepsilon}_{vit}^2}{(N - 1)(T - 1) - K} \tag{4.29}$$

where the $\hat{\varepsilon}_{vit}$ are residuals obtained from the regression of the deviations $y_{it} - \bar{y}_{i.} - \bar{y}_{.t} + \bar{y}_{..}$ on $x_{itk} - \bar{x}_{i.k} - \bar{x}_{.tk} + \bar{x}_{..k}$. These deviations are the elements of the Y_1 vector and W_1 matrix, respectively, defined for equation (4.4). The x_{itk} are values of the independent variables in the matrix W of equation (4.1). Here

$$\bar{y}_{i.} = \sum_{t=1}^{T} \frac{y_{it}}{T} \tag{4.30}$$

$$\bar{y}_{.t} = \sum_{i=1}^{N} \frac{y_{it}}{N} \tag{4.31}$$

$$\bar{y}_{..} = \sum_{i=1}^{N} \sum_{t=1}^{T} \frac{y_{it}}{NT} \tag{4.32}$$

$$\bar{x}_{i.k} = \sum_{t=1}^{T} \frac{x_{itk}}{N} \tag{4.33}$$

$$\bar{x}_{.tk} = \sum_{i=1}^{N} \frac{x_{itk}}{T} \tag{4.34}$$

$$\bar{x}_{..k} = \sum_{i=1}^{N} \sum_{t=1}^{T} \frac{x_{itk}}{NT} \tag{4.35}$$

and the v subscript on ε_{vit} is simply to distinguish this residual from residuals used in subsequent equations.

$$\hat{\sigma}_c^2 = \frac{\sum_{i=1}^{N} \sum_{t=1}^{T} \hat{\varepsilon}_{cit}^2 - [T(N-1) - K + 1]\hat{\sigma}_v^2}{T(N-1) - T\operatorname{tr}(U_1)} \tag{4.36}$$

where the $\hat{\varepsilon}_{cit}$ are residuals from the regression of the deviations $y_{it} - \bar{y}_{.t}$ on $x_{itk} - \bar{x}_{.tk}$,

$$U_1 = [W'(M_{12} + M_{1.})W]^{-1} W'M_{1.}W \tag{4.37}$$

$$M_{12} = I_{NT} - \frac{A}{T} - \frac{B}{N} + \frac{J}{NT} \tag{4.38}$$

$$M_{1.} = \frac{A}{T} - \frac{J}{NT} \tag{4.39}$$

A and B are as previously defined in equations (4.12) and (4.13) and J is an NT × NT matrix of ones.

$$\hat{\sigma}_s^2 = \frac{\sum_{i=1}^{N} \sum_{t=1}^{T} \hat{\varepsilon}_{sit}^2 - [N(T-1) - K + 1]\hat{\sigma}_v^2}{N(T-1) - Ntr(U_2)} \tag{4.40}$$

where the $\hat{\varepsilon}_{sit}$ are residuals from the regression of the deviations $y_{it} - \bar{y}_{i.}$ on $x_{itk} - \bar{x}_{i.k}$,

$$U_2 = [W'(M_{12} + M_{.2})W]^{-1}W'M_{.2}W \tag{4.41}$$

$$M_{.2} = \frac{B}{N} - \frac{J}{NT} \tag{4.42}$$

and the other terms have been previously defined.

Fuller and Battese show that these variance estimators are of the same probability order as those of Swamy and Arora.

Once estimators of the variance components have been chosen, these can be substituted into equation (4.14), using the expressions in (4.11) or (4.15), and the feasible Aitken estimator can be obtained.

$$\hat{d}_{EC} = (Z'\hat{\Omega}^{-1}Z)^{-1} Z'\hat{\Omega}^{-1}Y \tag{4.43}$$

Computationally, the inversion of the Ω matrix will not be efficient since this is an NT × NT matrix. Therefore, some transformation of the data is desirable. Fuller and Battese (1974) provide such a computational procedure for use with their estimates of the variance components. Once the estimates $\hat{\sigma}_v^2$, $\hat{\sigma}_c^2$ and $\hat{\sigma}_s^2$ are obtained, the GLS estimates are computed by the OLS regression of the dependent variables

$$y_{it} - \hat{\alpha}_1\bar{y}_{i.} - \hat{\alpha}_2\bar{y}_{.t} + \hat{\alpha}_3\bar{y}_{..} \tag{4.44}$$

on the independent variables

$$x_{itk} - \hat{\alpha}_1\bar{x}_{i.k} - \hat{\alpha}_2\bar{x}_{.tk} + \hat{\alpha}_3\bar{x}_{..k} \tag{4.45}$$

where

$$\hat{\alpha}_1 = 1 - \left(\frac{\hat{\sigma}_v^2}{\hat{\sigma}_v^2 + T\hat{\sigma}_c^2} \right)^{\frac{1}{2}} \tag{4.46}$$

$$\hat{\alpha}_2 = 1 - \left(\frac{\hat{\sigma}_v^2}{\hat{\sigma}_v^2 + N\hat{\sigma}_s^2} \right)^{\frac{1}{2}} \tag{4.47}$$

$$\hat{\alpha}_3 = \hat{\alpha}_1 + \hat{\alpha}_2 - 1 + \left(\frac{\hat{\sigma}_v^2}{\hat{\sigma}_v^2 + T\hat{\sigma}_c^2 + N\hat{\sigma}_s^2} \right)^{\frac{1}{2}} \tag{4.48}$$

This procedure can be applied whether the variance components are estimated using the Fuller and Battese estimators or other estimators of the variance components.

When the procedure of applying OLS to the transformed observations is used, the standard errors computed by the OLS regression program can serve as approximate standard errors for the estimated coefficients as noted by Fuller and Battese.

4.3 THE ERROR COMPONENTS MODEL WITH INDIVIDUAL COMPONENTS ONLY

The model in equation (4.1) has received extensive study under the assumption that the s_t are not present, that is, that the disturbances may be decomposed as

$$\varepsilon_{it} = c_i + v_{it} \tag{4.49}$$

where Assumptions 4.3 and 4.4 are maintained and the c_i and v_{it} are assumed to be independent.

The variance-covariance matrix for this model simplifies to

$$E(\varepsilon\varepsilon') = \Omega = \begin{bmatrix} \sigma_c^2 C_T & 0 & \cdots & 0 \\ 0 & \sigma_c^2 C_T & \cdots & 0 \\ \vdots & \vdots & & \vdots \\ 0 & 0 & \cdots & \sigma_c^2 C_T \end{bmatrix} \tag{4.50}$$

where C_T is defined in equation (4.10) and the zeros are T × T block matrices. The matrix Ω can also be written as

$$\Omega = \sigma_v^2 I_{NT} + \sigma_c^2 A \tag{4.51}$$

thus simplifying the form of equation (4.11).

The GLS estimator of d is

$$\tilde{d}_{EC} = (Z'\Omega^{-1}Z)^{-1}Z'\Omega^{-1}Y \tag{4.52}$$

where

$$\Omega^{-1} = \frac{1}{\sigma_v^2}\left(\gamma_1 A + I - \frac{A}{T}\right) \tag{4.53}$$

and

$$\gamma_1 = \frac{\sigma_v^2}{T(T\sigma_c^2 + \sigma_v^2)} \tag{4.54}$$

Following Arora (1973) the GLS estimator can be rewritten as

$$\tilde{d}_{EC} = \left(\frac{W_4' W_4}{\sigma_v^2} + \frac{W_3' N_3 W_3}{\sigma_3^2}\right)^{-1}\left(\frac{W_4' Y_4}{\sigma_v^2} + \frac{W_3' N_3 Y_3}{\sigma_3^2}\right) \tag{4.55}$$

where the elements of W_4 and Y_4 are computed by subtracting out the time series mean from each observation, the elements of W_3 and Y_3 are as described for equation (4.19), and $\sigma_3^2 = T\sigma_c^2 + \sigma_v^2$.

Again the operational form of such an estimator depends on how σ_3^2 and σ_v^2 (or, equivalently, σ_c^2 and σ_v^2) are estimated.

Arora (1973) and Swamy (1971, p. 35) suggest that σ_3^2 be estimated by using the MSE from the OLS regression of Y_3 on W_3 as in equation (4.25) and σ_v^2 be estimated using the MSE of the OLS regression of Y_4 on W_4:

$$\hat{\sigma}_v^2 = \frac{Y_4' M_4 Y_4}{N(T-1) - K + 1} \tag{4.56}$$

where

$$M_4 = I_{N(T-1)} - W_4(W_4'W_4)^{-1}W_4'$$

Nerlove (1971b) uses the variance component estimators in equations (4.26) and (4.27) when the time components are absent.

The MINQUE method of Rao (1972) can be used to estimate the variance components in this simpler model also. Wallace and Hussain's estimators (equations (4.20) to (4.22)) or Amemiya's estimators discussed previously could also be simplified to provide estimators of σ_c^2 and σ_v^2.

Harville (1977) discusses ML estimation of variance components in a general linear model that includes the error components model. See also Harville (1976).

Fuller and Battese (1973) suggest the following estimators:

Regress $y_{it} - \bar{y}_{i.}$ on the $x_{itk} - \bar{x}_{i.k}$ and use the residuals, $\hat{\varepsilon}_{vit}$, from this regression to construct the estimator

$$\hat{\sigma}_v^2 = \sum_{i=1}^{N} \sum_{t=1}^{T} \frac{\hat{\varepsilon}_{vit}^2}{N(T - 1) - K + 1} \tag{4.57}$$

Again, the x_{itk} represent the values of the independent variables of the matrix W of equation (4.1). Then, from the regression of Y in equation (4.1) on Z in equation (4.2), use the residuals $\hat{\varepsilon}_{cit}$ to estimate σ_c^2 as

$$\hat{\sigma}_c^2 = \frac{\sum_{i=1}^{N} \sum_{t=1}^{T} \hat{\varepsilon}_{cit}^2 - (NT - K)\hat{\sigma}_v^2}{NT - \mathrm{tr}(U)} \tag{4.58}$$

where

$$U = (Z'Z)^{-1} \sum_{i=1}^{N} T^2 \bar{z}_{i.}' \bar{z}_{i.} \tag{4.59}$$

and $\bar{z}_{i.}$ is the $1 \times K$ vector having kth element $\bar{z}_{i.k}$.

Once estimators of the variance components have been chosen, these can be substituted into equation (4.52), using the expression in (4.53), to construct a feasible Aitken estimator. Again, this procedure is inconvenient computationally. Fuller and Battese (1973) suggest regressing the transformed dependent variable

$$y_{it} - \hat{\alpha}_i \bar{y}_{i.} \tag{4.60}$$

on the transformed independent variables

$$x_{itk} - \hat{\alpha}_i \bar{x}_{i.k} \tag{4.61}$$

where

$$\hat{\alpha}_i = 1 - \left[\frac{\hat{\sigma}_v^2}{(\hat{\sigma}_v^2 + T\hat{\sigma}_c^2)} \right]^{\frac{1}{2}} \tag{4.62}$$

4.4 OTHER ERROR COMPONENTS MODEL DEVELOPMENTS

Taub (1979) showed that, in the error components model with individual effects or with both individual and time effects, the best linear unbiased predictor of $y_{i,T+S}$ is

$$y_{i,T+S}^0 = x_{i,T+S}^0 \tilde{d}_{EC} + (1 - \Theta)\bar{\tilde{e}} \tag{4.63}$$

where $\bar{\tilde{e}} = \bar{Y}_i - \bar{X}_i \tilde{d}_{EC}$ is the mean of the error components residuals for the ith cross-sectional unit with

$$\bar{Y}_i = \sum_{t=1}^{T} \frac{y_{it}}{T} \quad \text{and} \quad \bar{X}_i = \sum_{t=1}^{T} \frac{x_{it}}{T} \tag{4.64}$$

and

$$\Theta = \frac{\sigma_v^2}{(\sigma_v^2 + T\sigma_c^2)} \tag{4.65}$$

In practice, of course, estimates of σ_v^2 and σ_c^2 would be used in determining an estimator for Θ and $\hat{d}_{EC}$ (rather than $\tilde{d}_{EC}$) would be used in equation (4.63).

Hausman and Wise (1979) discuss attrition of sample members when using error components models and develop a maximum likelihood estimator of the model parameters. Biorn (1981) considers the case when some of the time series observations may be missing from the sample. The sampling plan he examines is a situation which might occur in a panel data study where, in consecutive time periods, certain individuals may be replaced by new individuals. Thus the sample size, N, remains the same but not all T time series observations are available for each individual. Biorn derives estimators for the regression coefficients and variance-covariance matrix and compares these to the complete sample estimators.

Hausman and Taylor (1981) examine the error components model with individual effects. They consider the model when the individual effects are correlated with some of the explanatory variables. They also allow time-invariant explanatory variables. An instrumental variable estimator is proposed that provides efficient estimates of the structural parameters. Amemiya and MaCurdy (1986) develop instrumental-variable estimators for error components models with different assumptions concerning the variance-covariance matrix and correlation of explanatory variables with the disturbances. They show that the Hausman and Taylor (1981) instrumental-variable estimator is inconsistent or asymptotically inefficient for certain of the model structures while proposing consistent and asymptotically efficient estimators in each case. See also Kiefer (1979) and Chamberlain (1978).

The error components model with serially correlated disturbances has been examined by Da Silva (1975) and Silver (1982). (See also Kiefer (1980) for an examination of the ANCOVA model with correlation over time.)

Mazodier and Trognon (1978) examine models with heteroskedastic error components. In their model the variances of the error components (either time or individual components) are allowed to vary within groups of individuals (or time periods).

In other studies of the error components model, Swamy and Mehta (1973a) examined the model with time and individual effects from a Bayesian viewpoint and Hussain (1969) studied the model with individual effects assumed to be parameters as in the ANCOVA model but with random time effects. Also in the model with individual effects only, Berzeg (1979) derived conditions for the existence of the ML estimator. However, he allowed the c_i and v_{it} to be correlated as did Kuh (1959). Griliches and Hausman (1986) examine the error components model with individual effects and errors in the independent variables. Hsiao (1985) also discusses this measurement problem.

Several authors have extended the error components approach to the SUR model of Chapter 3.

Consider the EC model written as in equation (4.1):

$$Y = b_0 \iota_{NT} + Wd + \varepsilon \tag{4.66}$$

$$= Z\bar{\beta} + \varepsilon \tag{4.67}$$

where the elements of ε are decomposed as in equation (4.3):

$$\varepsilon_{it} = c_i + s_t + v_{it} \tag{4.68}$$

with Assumptions 4.3, 4.4, 4.5 and 4.6 satisfied. Now extend this model to a case where M equations of the form (4.67) exist:

$$\begin{aligned} Y_1 &= Z_1\bar{\beta}_1 + \varepsilon_1 \\ &\;\vdots \\ Y_M &= Z_M\bar{\beta}_M + \varepsilon_M \end{aligned} \tag{4.69}$$

where each Y_j ($j = 1,\ldots,M$) is an NT $\times$ 1 vector of observations on the dependent variable, each Z_j contains NT observations on each independent variable, the ε_j are NT $\times$ 1 vectors of disturbances and the $\bar{\beta}_j$ are the K $\times$ 1 vectors of coefficients to be estimated. (The number of independent variables need not be the same in each equation so K could be changed to K_j.)

As noted by Avery (1977), each equation in (4.69) could be estimated separately if

$$E(\varepsilon_j\varepsilon_{j'}) = 0 \text{ for } j \neq j' \tag{4.70}$$

However, if the disturbance vectors are correlated across equations, a more efficient estimator can be developed. Avery develops these estimators by extending the single equation methods discussed earlier for SUR and EC estimation.

Baltagi (1980) uses the results in Amemiya (1971) to develop an alternative estimator to the one suggested by Avery. Prucha (1984) shows that the above two estimators are members of a class of asymptotically efficient estimators of the slope coefficients. Verbon (1980) applied a slightly generalized version of Avery's (1977) model as did Beierlein, Dunn, and McConnon (1981) and Salvas-Bronsard (1978). Kang (1983) applied the SUR model with error components to model demand for goods in different cities over time. Also considered are cases when the individual and time effects are fixed and when one effect is fixed but the other random.

Schmidt (1983) considers the ANCOVA model with individual effects only and contemporaneously correlated disturbances. He shows that SUR estimation with individual dummy variables is equivalent to SUR estimation on data transformed by the "within" transformation. This is not the case for arbitrary disturbance covariance structures (see, for example, Kiefer, 1980) but does work in the SUR case. Conniffe (1982b) also examines relationships between ANCOVA and SUR models.

Bhargava and Sargan (1983) consider the model with individual error components only and view it as a system of simultaneous equations. They allow a lagged value of the dependent variable to be included in the equation as an independent variable. Assuming either 1) the y_0 are exogenous or 2) the y_0 are endogenous and explained by a particular structure, they show how to estimate the model using limited information maximum likelihood (LIML) methods. Their simulation results suggest that when the y_0 are endogenous the LIML estimation procedure performs extremely well. They also develop a model which allows the disturbances and the regressors to be correlated and generalize the LIML procedure in this case.

The error components model has also been extended to systems of simultaneous equations. See, for example, Parhizgari and Davis (1978), Baltagi (1981b, 1984), and Prucha (1985). Magnus (1982) extends the error components model to multivariate regression and analyzes both linear and nonlinear models. His model is applied in Sickles (1985).

4.5 ANALYTIC RESULTS

Since a number of estimators for the variance components in the error components model are available, the researcher is left with a choice of which to use. A question to be asked is whether there is one estimator that is "better" than the others according to some criterion.

Baltagi (1981a) suggests three criteria to help in choosing appropriate estimators. The criteria are 1) the simplicity and practicality of the estimator, 2) analytic or exact finite sample results and 3) Monte Carlo studies of small sample performance of estimators. In this section the analytic or exact finite sample results are examined.

The small sample properties of estimators in the error component model are especially important since Swamy and Arora (1972) and Swamy (1971, pp. 36) have shown that there are an infinite number of asymptotically efficient estimators of d, the vector of slope coefficients. Thus it is not possible to choose an estimator on the basis of asymptotic properties alone. Included in the class of asymptotically efficient estimators described by Swamy and

Arora are their own estimator as previously described and the ANCOVA estimator discussed in the first section of this chapter. Since asymptotics do not provide a clear choice of an estimator, the exact finite sample results may help. (Note that the result in Don and Magnus (1980) can be used to show unbiasedness under certain conditions.)

For the error components model with time and individual components, Swamy and Arora (1972) studied the small sample properties of their estimator. They showed that it may be less efficient than CP (classical pooling) if N and T and the true values of $\sigma_c{}^2$ and $\sigma_s{}^2$ are small. It may also be less efficient than the ANCOVA estimator if N and T are small and the true values of $\sigma_c{}^2$ and $\sigma_s{}^2$ are large. Otherwise, the GLS estimator using the estimates of the variance components in equations (4.23), (4.24), and (4.25) should be relatively more efficient than CP and ANCOVA. As guidelines, Swamy and Arora suggest that $N - K$ and $T - K$ (where K is the number of explanatory variables) both be larger than ten before the GLS estimator is considered.

Under the assumption that the time effects are absent, Swamy and Mehta (1979) have reduced the choice of estimators to a choice among six possible estimators when the disturbances are normal. The choice among estimators depends on a priori information concerning relative magnitudes of error variances.

Taylor (1980) also examined the model with individual components. His results suggest that the only ambiguity in choice of an estimator will be when $N - K$ is less than or equal to ten. In cases where $N - K > 10$, generalized least squares using the estimated variance components will be more efficient than CP or ANCOVA.

Anderson and Hsiao (1981) analyzed the error components model with individual effects and a lagged value of the dependent variable as an explanatory variable. The model they examined can be written as

$$y_{it} = \beta y_{i,t-1} + c_i + v_{it} \tag{4.71}$$

They study estimators under four different sets of initial conditions:

Case 1. y_{i0} are observed fixed constants.

Case 2. y_{i0} are random with a common mean.

Equation (4.71) may be written to allow dependence between c_i and y_{i0} as

$$y_{it} = \eta_{it} + \gamma_i \qquad t = 0, 1, \ldots, T \tag{4.72}$$

$$v_{it} = \beta\eta_{i,t-1} + u_{it} \qquad t = 1, \ldots, T \tag{4.73}$$

where η_{it}, γ_i and u_{it} are unobservable.

Case 3. η_{i0} are fixed.

Case 4. η_{i0} are random.

They show that the ML estimator is inconsistent in Case 3 when T is fixed and $N \to \infty$. The ANCOVA estimator is inconsistent in all cases when T is fixed and $N \to \infty$.

A new instrumental variable estimator similar to Nerlove's (1971b) is proposed. They suggest using one of the following estimators:

$$\hat{\beta}_{AH1} = \frac{\sum_{i=1}^{N}\sum_{t=3}^{T}(y_{it} - y_{i,t-1})(y_{i,t-2} - y_{i,t-3})}{\sum_{i=1}^{N}\sum_{t=3}^{T}(y_{i,t-1} - y_{i,t-2})(y_{i,t-2} - y_{i,t-3})} \tag{4.74}$$

if there is prior belief that successive observations are positively correlated, and

$$\hat{\beta}_{AH2} = \frac{\sum_{i=1}^{N}\sum_{t=2}^{T}(y_{it} - y_{i,t-1})\, y_{i,t-2}}{\sum_{i=1}^{N}\sum_{t=2}^{T}(y_{i,t-1} - y_{i,t-2})\, y_{i,t-2}} \tag{4.75}$$

if succesive observations are thought to be negatively correlated.

The estimators $\hat{\beta}_{AH1}$ and $\hat{\beta}_{AH2}$ are shown to be consistent whenever $N \to \infty$ or $T \to \infty$ or both.

Anderson and Hsiao (1982) also study the model when there are exogenous variables. They suggest the use of the same estimators for β. If the equation is written as

$$y_{it} = \beta y_{i,t-1} + \delta' x_i + c_i + v_{it} \tag{4.76}$$

where x_i is a $K_1 \times 1$ vector of values on time invariant explanatory variables and δ' is a $1 \times K_1$ vector of coefficients, then the IV estimator of β will be consistent when $N \to \infty$ or $T \to \infty$ or both. To estimate δ, substitute $\hat{\beta}_{AH1}$ or $\hat{\beta}_{AH2}$ into the equation

$$\bar{y}_i - \beta\bar{y}_{i,-1} = \delta' x_i + c_i + v_i \tag{4.77}$$

and apply least squares. The LS estimator of δ will be consistent if T is fixed and $N \to \infty$ but inconsistent if N is fixed and $T \to \infty$ (as will the ML estimator of δ; the parameter δ is not estimable using ANCOVA). See Nickell (1981) for a related article.

When time varying exogenous variables are present the model may be written in one of two forms:

$$y_{it} = \beta y_{i,t-1} + \gamma' x_{it} - \beta\gamma' x_{i,t-1} + c_i + v_{it} \tag{4.78}$$

or

$$y_{it} = \beta y_{i,t-1} + \gamma' x_{it} + c_i + v_{it} \tag{4.79}$$

where x_{it} is a $K_1 \times 1$ vector of time varying exogenous variables and γ is a $K_1 \times 1$ vector of coefficients to be estimated.

The first model form has been called the serial correlation model and the second the state dependence model. See Heckman (1978). In the serial correlation model y_{it} is affected only by x_{it}; in the state dependence model y_{it} is affected by x_{it}, $x_{i,t-1}, \ldots$. Anderson and Hsiao (1982, p.61) further describe the difference as follows:

> [In the serial correlation model]...if x is increased in period t and then returned to its former level, the distribution of y in period t + 1 is not affected. Past y is informative because it helps to predict the effect of unobservable variables which are serially correlated. In the state dependence model a change in x that affects the distribution of y in period t will continue to affect the distribution of y in period t + 1, even though the intervention is present only in period t.

A similar instrumental variable approach is suggested in these cases. For alternative approaches to the treatment of models such as those in equations (4.78) and (4.79) see Chamberlain (1982, 1984). For further analytical results see Trognon (1978).

4.6 SIMULATION RESULTS

Baltagi (1981a) used a Monte Carlo simulation to compare alternative estimators for the error components model with both individual and time effects. The estimators compared included the CP estimator, ANCOVA, and the two stage GLS procedures developed by Wallace and Hussain (1969), Amemiya (1971), Swamy and Arora (1972), Rao (1972), Fuller and Battese (1974) and Nerlove (1971b). Each of these procedures was described in Section 4.2 of this chapter.

The simulation used N = 25 cross sections and T = 10 time series observations. The model examined was

$$y_{it} = 5 + 0.5x_{it} + c_i + s_t + v_{it} \tag{4.80}$$

One hundred Monte Carlo trials were used for each factor combination in the experiment. The factors were determined by varying σ_c^2 and σ_s^2 over the range 0.0, 0.2, 4.0, 8.0, 12.0 and 16.0 where $\sigma_c^2 + \sigma_s^2 + \sigma_v^2$ was always set equal to 20.0 and $1 - \sigma_c^2/(\sigma_c^2 + \sigma_s^2 + \sigma_v^2) - \sigma_s^2/(\sigma_c^2 + \sigma_s^2 + \sigma_v^2)$ was always positive.

Baltagi found that there were always gains in efficiency using the two-stage methods rather than OLS or ANCOVA unless the variance components were very close to zero. However, all the two-stage methods performed reasonably well (according to the relative mean squared error criteria) leading to no clear-cut best choice. The results also indicated that better estimates of the variance components do not necessarily lead to better coefficient estimates and that replacing negative variance estimates by zero did not have a serious effect on the performance of the second-round GLS estimates of the coefficients.

One further recommendation was to use more than one of the two-stage GLS procedures and compare the estimates from the alternate procedures. If the estimates differed widely, the model specification would be questioned and further investigated.

On the basis of Baltagi's results one might rule out the MINQUE procedure of Rao (1972) due to its increased computational complexity. Also, Wallace and Hussain's estimator is not consistent when lagged values of the dependent variable appear as explanatory variables. Baltagi's experiment did not include stochastic regressors, however, so it is unclear which of the remaining GLS estimators will perform well in this case.

When time components are assumed absent there are several Monte Carlo experiments to consider.

Nerlove (1971b) examined the performance of several estimators using the model

$$y_{it} = \beta_1 y_{i,t-1} + \beta_2 x_{it} + c_i + v_{it} \tag{4.81}$$

The exogenous variable x_{it} was generated as

$$x_{it} = 0.1t + 0.5x_{i,t-1} + \gamma_{it} \tag{4.82}$$

where γ_{it} was chosen as uniformly distributed on the interval $[-\frac{1}{2}, \frac{1}{2}]$ and x_{i0} was chosen as $5 + 10\ \gamma_{i0}$. Sample sizes used were N = 25 and T = 10. Parameter values used were β_1 = 0.1, 0.3, 0.5, 0.7, and 0.9; β_2 = 0.0, 0.5 and 1.0; $\sigma_c^2/(\sigma_c^2 + \sigma_v^2)$ = 0.0, 0.15, 0.3, 0.45, 0.6, 0.75, 0.9, and 0.95, with $\sigma_c^2 + \sigma_v^2$ always set to equal 1. Fifty Monte Carlo trials for each factor combination were used.

Estimators examined included CP, ANCOVA, the Nerlove (1971b) estimator described previously, and the ML estimator. Nerlove's results favored the feasible Aitken estimator of Nerlove (1971b). The estimator compared favorably with the others in terms of both bias and mean squared error.

Nerlove (1967) also examined the model in (4.81) without the exogenous variable:

$$y_{it} = \beta_1 y_{i,t-1} + c_i + v_{it} \tag{4.83}$$

for sample sizes N = 25 and T = 10.

The design of the simulation was similar to that for the model in equation (4.81). The feasible Aitken estimator of Nerlove (1971b) again performed favorably when compared to CP, ANCOVA, and ML.

Arora (1973) compared the estimator in equation (4.55) using his suggested estimators of the variance components to CP, the OLS estimator applied to data aggregated over time, and ANCOVA. The model examined was

$$y_{it} = \beta_0 + \beta_1 x_{it} + c_i + v_{it} \tag{4.84}$$

The variable x_{it} was generated as

$$x_{it} = 0.1(t - 1) + 1.05x_{i,t-1} + \gamma_t \tag{4.85}$$

where the γ_t were chosen as uniformly distributed in the range 0 to 1. Initial values of x_{i0} were chosen as uniformly distributed numbers in the range 0 to 100. Parameter values used were $\beta_0 = 0$ and 5; $\beta_1 = 0.5$ and 0.8; $\sigma_c^2/(\sigma_c^2 + \sigma_v^2) = 0.0$, 0.4, and 0.8, with $\sigma_c^2 + \sigma_v^2$ set equal to 10. Sample sizes were N = 25 and T = 6. The estimator in equation (4.55) performed well relative to the other estimators.

Maddala and Mount (1973) also examined the model in equation (4.84). For sample size N = 25 and T = 10 the x_{it} values used by Nerlove (1971b) from equation (4.82) were used. The parameter values used were $\beta_0 = 0.0$; $\beta_1 = 1.0$; $\sigma_c^2/(\sigma_c^2 + \sigma_v^2) =$ 0.5, 0.11, and 0.002. The estimators examined included CP, ANCOVA, ML, Wallace and Hussain's estimator, Amemiya's estimator, Nerlove's estimator, an estimator similar to that of Fuller and Battese, and the MINQUE estimator.

Results of the simulation showed that, for large values of σ_c^2, any of the GLS estimators outperformed the CP estimator and the ANCOVA estimator. There was little difference between the GLS estimators, however.

Baltagi (1986) compared the cross-sectionally heteroskedastic and timewise autoregressive structure of Kmenta (1971, pp. 508-514) and the error components model under sets of assumptions appropriate to each. (See Chapter 2 for a discussion of the Kmenta model.) Baltagi concludes that for T small and N large, it is more serious to apply the Kmenta technique when disturbances have an error components structure than to apply EC techniques when disturbances are of the Kmenta type.

Baltagi and Griffin (1984) provide some additional suggestions on estimation of short and long run effects.

4.7 INFERENCE IN ERROR COMPONENTS MODELS

The covariance matrix of the EC estimator is given by

$$(Z'\hat{\Omega}^{-1}Z)^{-1} \tag{4.86}$$

from equation (4.14). The standard errors of the regression coefficients can be computed as the square roots of the diagonal elements of this matrix. These can be used to construct t-statistics to test hypotheses about the population regression coefficients. Alternatively, Fuller and Battese (1973, 1974) have suggested a transformation of the data after which least squares can be applied to produce EC estimates of the structural coefficients. The standard errors of the estimated coefficients should be well approximated by the least squares standard error. See equations (4.44) through (4.48) or (4.60) through (4.62) for the appropriate transformations.

Breusch and Pagan (1980) provide a test of the hypothesis

$$H_0: \sigma_c^2 = \sigma_s^2 = 0 \tag{4.87}$$

If the null hypothesis is accepted, it can be assumed that both the individual and time components are absent from the model. The model reduces to the CP model of Chapter 2. Their test is based on the Lagrange multiplier (LM) method and the resulting test statistic is given by

$$\lambda_{LM1} = \frac{NT}{2}\left\{\frac{1}{T-1}\left[\frac{\sum_{i=1}^{N}\left(\sum_{t=1}^{T}\hat{\varepsilon}_{it}\right)^2}{\sum_{i=1}^{N}\sum_{t=1}^{T}\hat{\varepsilon}_{it}^2} - 1\right]^2 + \frac{1}{N-1}\left[\frac{\sum_{t=1}^{T}\left(\sum_{i=1}^{N}\hat{\varepsilon}_{it}\right)^2}{\sum_{i=1}^{N}\sum_{t=1}^{T}\hat{\varepsilon}_{it}^2} - 1\right]^2\right\} \tag{4.88}$$

where the $\hat{\varepsilon}_{it}$ are the residuals obtained from applying least squares to equation (4.1).

Under the null hypothesis, λ_{LM1} will have a chi-square distribution with two degrees of freedom.

To test whether the individual components alone are absent

$$H_0: \sigma_c^2 = 0 \tag{4.89}$$

the test statistic will be

$$\lambda_{LM2} = \frac{NT}{2(T-1)}\left[\frac{\sum_{i=1}^{N}\left(\sum_{t=1}^{T}\hat{\varepsilon}_{it}\right)^2}{\sum_{i=1}^{N}\sum_{t=1}^{T}\hat{\varepsilon}_{it}^2} - 1\right]^2 \tag{4.90}$$

Under H_0, λ_{LM2} will have a chi-square distribution with one degree of freedom.

One consideration in choosing between the ANCOVA and EC models is whether the error components and the independent variables are uncorrelated. If they are uncorrelated, the EC estimator is appropriate (one of the two-step generalized least squares estimators); however, if the error components and the independent variables are correlated, the ANCOVA estimator should be used. The ANCOVA estimator will be best linear unbiased in such a case, conditional on the c_i and/or s_t in the sample. The EC estimator will be biased.

Hausman (1978) developed a test of the hypothesis:

H_0: no correlation exists between the independent variable and the error components

The test statistic is

$$M = (\hat{d}_{COV} - \hat{d}_{EC})[V(\hat{d}_{COV}) - V(\hat{d}_{EC})]^{-1}(\hat{d}_{COV} - \hat{d}_{EC}) \quad (4.91)$$

where $\hat{d}_{COV}$ is the ANCOVA estimator, $\hat{d}_{EC}$ is any of the EC estimators, $V(\hat{d}_{COV})$ is the variance-covariance matrix of the ANCOVA estimator, and $V(\hat{d}_{EC})$ is the variance-covariance matrix of the EC estimator. The statistic applies whether individual and time components or individual components alone are present with appropriate computational adjustments.

Under the null hypothesis, M will have a chi-square distribution with K − 1 degrees of freedom (K − 1 = number of independent variables excluding the constant).

If the null hypthesis is rejected, the use of the ANCOVA estimator is supported. However, Hausman (1978) cautions that the ANCOVA estimator will be more severely affected by errors in variables than will the EC estimator so this tradeoff must be considered.

Another test for correlation between the independent variables and the error components is suggested by Chamberlain (1982, 1984). Chamberlain's approach is designed particularly for use with data sets involving a large number of individuals but short time series length.

Baltagi (1981a) examined the performance of the Hausman test in a Monte Carlo simulation. The test performed well in the sense of having a low frequency of committing a Type I error. The power of the test was not investigated.

Baltagi also examined the Breusch and Pagan (1980) test in the simulation experiment. He found that the LM test statistics performed well when the variance components were large. That is, with nonzero variance components the test tended to reject the null hypotheses of either equations (4.87) or (4.89). However, when variance components were set equal to zero (or very small) the test also tended to reject the null hypothesis in a large percentage of the trials. Unfortunately these are the cases when negative variance estimates often occur and a reliable test for zero variance components is important for proper model specification. When negative variance estimates do occur, regardless of the outcome of the LM test, Baltagi suggested replacing the negative estimates by zero. This substitution did not adversely affect the two-stage GLS estimates of the coefficients.

4.8 APPLICATIONS

The error components model has been used in a wide variety of applications. A few of these applications will be discussed here to illustrate the possible uses of the model.

The model in equations (4.1) to (4.3) with Assumptions 4.3 to 4.6 has been applied in the following situations. This is the model specification when both individual and time components are assumed to be present. Archibald and Gillingham (1980) model the short run demand for gasoline and use the Fuller and Battese (1974) estimators of the variance components. Karathanassis and Tzoannos (1977) examine the ability of alternative monetary theories to explain the demand for money by business firms. They apparently use the Wallace and Hussain (1969) estimators. Wilbur, Miller, and Brown (1985) use TSCSREG, a procedure developed by SAS (1982), to estimate rates of return to assets financed by means of earnings retentions, new equity, and new debt in the commercial banking industry. Lee and Chang (1986) examine the importance of dividend policy on firm return.

In a financial application, Chang and Lee (1977) examine the impact on price per share of changes in dividends and retained earnings. They apply both the ANCOVA and EC estimation procedures and compare results including both individual and time effects with models involving only time effects and models with only individual effects. Their results indicate that more variation in price is explained when both time and individual effects are included.

With time effects assumed absent, Ward and Davis (1978) study the effect of coupon availability on product demand. They use a ML estimator similar to the procedure described in Maddala (1977, p. 345) to estimate the model. Parsons (1974) uses a dynamic model to examine whether consumer advertising increases retail availability of a new product. He uses Nerlove's (1971b) estimation procedure. Johnson and Oksanen (1977) study the demand for alcoholic beverages in Canada again using the Nerlove (1971b) estimation procedure. Moriarty (1975) estimates the relationship between market share of products and a number of independent variables where the individual effects correspond to sales districts.

Houthakker, Verleger, and Sheehan (1974) used the individual error components model to examine the demand for gasoline and electricity. The Balestra and Nerlove (1966) estimation procedure was used since a lagged dependent variable was used as an explanatory variable. See also Kwast (1980) and Mehta, Narasimham, and Swamy (1978). Berzeg (1982) extended the analysis of Houthakker, Verleger, and Sheehan (1974) by using a generalized error components approach. Berzeg allows the individual error component, c_i, and the disturbance, v_{it}, in equation (4.49) to be correlated as suggested by Kuh (1959). The maximum likelihood method proposed by Berzeg (1979) was used to estimate the model.

Park, Mitchell, Wetzel, and Alleman (1983) use a model with individual error components to study the household demand for telephone calls before and after introduction of measured rates. The stochastic portion of their model is actually somewhat more complicated than the decomposition in equation (4.49). Their model contains an individual error component, c_i, specific to each household and an additional stochastic component that applies only to the time periods when measured service tariffs are in effect. They estimate the model by maximum likelihood using LISREL, a computer program due to Joreskog and Sorbom (1978). See also Joreskog (1978).

Weiss and Lillard (1978) and Lillard and Weiss (1979) apply a more general form of error components model to examine the relationship between earnings of scientists and several explanatory variables (see also Lillard, 1978). They decompose the disturbance in the model as

$$\varepsilon_{it} = c_{1i} + c_{2i}(t - \bar{t}) + u_{it} \tag{4.92}$$

where $u_{it} = \rho u_{i,t-1} + v_{it}$. The c_{1i} represent the unobservable individual effects on earnings. The c_{2i} represent the effect of omitted variables which affect the growth of earnings. The c_{1i} and c_{2i} are allowed to be correlated. The remaining disturbances, u_{it}, are serially correlated but the serial correlation coefficient is the same for each cross section. They also use the LISREL ML estimation procedure. See MaCurdy (1982) for discussion of a general model which encompasses many of these error covariance structures.

Baltagi and Griffin (1983) apply several of the two-step GLS procedures discussed in this chapter to estimate a model with individual effects. They discuss why different procedures might produce estimates which differ considerably in certain instances.

For other applications see Nielsen and Hannan (1977), Nerlove (1965), Mairesse (1978), Henin (1978), Chang (1978), Vernon and McElroy (1973), Margolis (1982), Schmidt and Sickles (1984), Render and Neumann (1980), and Owusu-Gyapong (1986).

A survey of many of the current issues in error components models is presented in Hsiao (1985).

As noted in this chapter, ANCOVA and error components estimation can be performed by using OLS applied to appropriately transformed data. Analysis of covariance routines are also available in statistical packages such as SPSSX (1986). ANCOVA estimates can also be obtained by applying OLS to equations with dummy variables inserted to represent firm and/or time effects although this procedure is not computationally efficient.

Several programs are available which are designed to compute estimates of error components model parameters. Freiden (1973) developed a program using Nerlove's (1971b) estimator for error components models with individual effects. The program is designed to handle lagged dependent variables used as explanatory variables. Henry, McDonald, and Stokes (1976) modified Freiden's program to allow explanatory variables which are constant over time. Their procedure has been implemented in a program discussed by Stokes (1981).

Joreskog (1978) discusses the use of the LISREL program to estimate models with a generalized error component structure.

Drummond and Gallant (1977) have developed several procedures which are available from SAS (1982). The Fuller and Battese (1974) method, allowing for individual and/or time effects, has been implemented, as well as the methods due to Da Silva (1975) for models with error components and moving average errors.

Hall (1978) discusses the package TSCS which contains procedures for estimating several of the models discussed throughout this book. The ANCOVA model with individual effects and the error components model with individual effects are included. This package is distributed by TSP International.

4.9 EXAMPLE: INVESTMENT FUNCTION ESTIMATION

The data discussed in Section 2.3 will be used again in this section to illustrate the ANCOVA and EC procedures. The equation to be estimated is

$$Y_{it} = \beta_0 + \beta_1 X_{1i,t-1} + \beta_2 X_{2i,t-1} + \varepsilon_{it} \tag{4.93}$$

where Y_{it}, $X_{1i,t-1}$ and $X_{2i,t-1}$ are as defined in Section 2.3. The coefficients β_1 and β_2 are assumed to be equal for each cross-sectional unit as is the overall intercept β_0. For illustrative purposes, assume the disturbance, ε_{it}, can be decomposed as

$$\varepsilon_{it} = c_i + v_{it} \tag{4.94}$$

where c_i represent time-invariant individual effects and the v_{it} are the remaining random effects.

If the effects, c_i, are viewed as fixed parameters to be estimated, then ANCOVA is the appropriate estimation procedure. The GLM procedure in SAS was used to produce the ANCOVA estimates. These are shown in the first row of Table 4.1 along with t-statistics in parentheses. The SAS procedure assumes there is an overall constant, β_0, to be estimated. One of the individual effects is, therefore, not estimable and its value is set equal to zero. In this case the effect for the tenth cross-

TABLE 4.1 Estimates of Investment Function Coefficients Using ANCOVA and EC Estimation

	$\hat{\beta}_0$	$\hat{\beta}_1$	$\hat{\beta}_2$
ANCOVA	-6.6	0.110	0.331
	(-0.55)	(9.26)	(17.88)
EC	-50.5	0.111	0.274
	(-4.67)	(17.12)	(12.80)

TABLE 4.2 Individual Effect Estimates for ANCOVA Estimation

$\hat{c}_1$	-62.6	$\hat{c}_6$	-16.5
$\hat{c}_2$	107.4	$\hat{c}_7$	-60.1
$\hat{c}_3$	-228.6	$\hat{c}_8$	-50.8
$\hat{c}_4$	-21.1	$\hat{c}_9$	-80.7
$\hat{c}_5$	-108.8		

sectional unit was equal to zero. It essentially serves as a base level unit from which to measure the other individual effects. The individual effects estimates are shown in Table 4.2.

When the c_i are assumed to be random the EC procedure is appropriate. The Swamy and Arora procedure was used to estimate the variance components. The estimate of the overall variance, σ_V^2, was 2781.7 while the individual component variance, σ_C^2, was estimated as 221.9. The resulting coefficient estimates are shown in the second row of Table 4.1 along with t-statistics in parentheses.

5

Random Coefficient Regression

5.1 THE RANDOM COEFFICIENT REGRESSION MODEL

The error components model provides an alternative to classical pooling and is an improvement over the ANCOVA model because of the reduction in the number of parameters to be estimated. Unfortunately the use of the error components model has not alleviated one especially restrictive assumption: It was assumed that the slope coefficients, d (see equation 4.1), for each cross-sectional unit were equal. This may be an acceptable assumption in certain applications, but it is often violated. We are thus led to search for an approach that allows the coefficients of interest to differ but provides some method of modeling the cross-sectional units as a group rather than individually.

One possibility would be to introduce into the model dummy variables that would indicate differences in the coefficients across individual units, that is, develop an approach similar to the ANCOVA model (Maddala 1977, p.321, eq. 14.4). Such a dummy variable approach would prove unwieldy if there were many coefficients or many individual units, however, owing to the large number of parameters to be estimated. Alternatively, the error components model could be generalized still further, treating not just the intercept but all the coefficients as random. This is the approach that will be taken in this chapter.

Consider again the model in equation (1.8):

$$y_i = X_i\beta_i + \varepsilon_i \tag{5.1}$$

for i = 1,...,N.

Suppose each regression coefficient is now viewed as a random variable; that is, the coefficients, β_i, are viewed as invariate over time, but varying from one unit to another. Particular distributions for these random variables can also be assumed for purposes of inference.

The RCR model has been justified in various applications. For example, Mehta, Narasimham, and Swamy (1978) use the RCR model to examine the demand for gasoline. The coefficient vectors, β_i, represent the effects of various explanatory variables on demand for different states. Mehta et al. note that since the β_i are coefficient vectors of the same economic relationship they very well could be related. If they are near $\bar{\beta}$, then they are related through this proximity and the relationship may be expressed as a probability distribution with a specified mean and variance-covariance matrix, thus leading to the RCR model.

Kraft and Rodekohr (1978), again examining gasoline demand functions, justify use of the RCR model by suggesting that variation between coefficients is "due to the infrastructure and mass transit possibilities which affect the consumers' alternatives when faced with changes in gasoline prices and income."

Ferguson and Leech (1978) used the RCR model to estimate a yield function for a pine tree plantation. The cross-sectional units in their study were plots. Since each plot was regarded as one item in a random sample from the population of all plots, the coefficients could be regarded as random variables.

Rewrite the coefficients β_i as

$$\beta_i = \bar{\beta} + v_i \tag{5.2}$$

where $\bar{\beta}$ is a fixed component, the common mean of the distribution from which each coefficient is drawn, and the v_i are random components each with mean zero. The v_i allow the coefficients to differ from unit to unit.

Using the notation in (5.1) and (5.2) the model can be written as

$$Y_i = X_i(\bar{\beta} + v_i) + \varepsilon_i = X_i\bar{\beta} + e_i \tag{5.3}$$

where $e_i = X_i v_i + \varepsilon_i$. As in the error components model the random components are combined into a single disturbance term, e_i.

Again, there is a nonspherical disturbance covariance matrix and a generalized least squares estimator will provide a more efficient estimator of $\bar{\beta}$ than a technique such as classical pooling.

The model in (5.1) or (5.3) is the random coefficient regression (RCR) model examined by Swamy in several publications (Swamy 1970, 1971, 1973, and 1974). The statistical estimation procedure he developed is a two-step procedure. First, separate regressions are calculated for each cross-sectional unit. Then the population mean and variance for the distribution of each regression coefficient are estimated. The estimates of the population means are a weighted average of the OLS estimates of each individual coefficient. The weights used involve both the standard error of each regression and the covariance structure of the random coefficients. The variances of the coefficients are estimated as a sample variance of the individually estimated coefficients allowing for variation arising from the residuals, ε_i, in the model.

Equation (5.3) applies to each of N cross-sectional units. These N equations can be rewritten as

$$Y = X\bar{\beta} + e \tag{5.4}$$

where

$$Y = \begin{bmatrix} Y_1 \\ Y_2 \\ \cdot \\ \cdot \\ \cdot \\ Y_N \end{bmatrix} \quad X = \begin{bmatrix} X_1 & 0 & \cdots & 0 \\ 0 & X_2 & \cdots & 0 \\ \cdot & \cdot & & \cdot \\ \cdot & \cdot & & \cdot \\ \cdot & \cdot & & \cdot \\ 0 & 0 & \cdots & X_N \end{bmatrix} \quad \text{and} \quad e = \begin{bmatrix} e_1 \\ e_2 \\ \cdot \\ \cdot \\ \cdot \\ e_N \end{bmatrix} \tag{5.5}$$

The following assumptions are made:

Assumption 5.1: The sample sizes are such that $N > K$ and $T > K$.

Assumption 5.2: The independent variables are nonstochastic in the sense that X_i is fixed in repeated samples on Y_i. The rank of X is K.

Assumption 5.3: The ε_i are independently and identically distributed with $E(\varepsilon_i) = 0$ and $E(\varepsilon_i \varepsilon_i') = \sigma_i^2 I_T$.

Assumption 5.4: The coefficient vectors β_i are independently and identically distributed with $E(\beta_i) = \bar{\beta}$ and $E(\beta_i - \bar{\beta}) \times (\beta_i - \bar{\beta})' = \Delta$.

Assumption 5.5: The ε_i and β_j are independent for every i and j.

The variance-covariance matrix of e is

$$E(ee') = \Omega = \begin{bmatrix} X_1 \Delta X_1' + \sigma_1^2 I_T & 0 & \cdots & 0 \\ 0 & X_2 \Delta X_2' + \sigma_2^2 I_T & \cdots & 0 \\ \vdots & \vdots & & \vdots \\ 0 & 0 & \cdots & X_N \Delta X_N' + \sigma_N^2 I_T \end{bmatrix} \tag{5.6}$$

where the zeros are all $T \times T$ null matrices and Δ is the variance-covariance matrix of the β_i as given in Assumption 5.4.

The GLS estimator of $\bar{\beta}$ is

$$\tilde{\bar{\beta}} = (X'\Omega^{-1}X)^{-1} \; X'\Omega^{-1}Y \tag{5.7}$$

which is shown by Swamy (1970, p.101) to be equivalent to

$$\tilde{\bar{\beta}} = \left\{ \sum_{i=1}^{N} [\Delta + \sigma_i^2 (X_i'X_i)^{-1}]^{-1} \right\}^{-1} \sum_{i=1}^{N} [\Delta + \sigma_i^2 (X_i'X_i)^{-1}]^{-1} \hat{\beta}_i \tag{5.8}$$

where $\hat{\beta}_i = (X_i'X_i)^{-1} X_i'Y_i$ is the OLS estimator of β_i.

The estimator, $\tilde{\bar{\beta}}$, in equation (5.8) can be viewed as the weighted average of the OLS estimators, $\hat{\beta}_i$, with weights, W_i, equal to

$$W_i = \left\{ \sum_{j=1}^{N} [\Delta + \sigma_j^2 (X_j'X_j)^{-1}]^{-1} \right\}^{-1} [\Delta + \sigma_i^2 (X_i'X_i)^{-1}]^{-1} \tag{5.9}$$

Of course, the GLS estimator cannot be used in practice since Δ and the σ_i^2 are unknown. Swamy (1970, p. 107) suggests the following unbiased and consistent estimators:

$$\hat{\sigma}_i^2 = \frac{\hat{\varepsilon}_i'\hat{\varepsilon}_i}{T - K} \tag{5.10}$$

which is simply the mean squared error from the OLS regression of Y_i on X_i;

$$\hat{\Delta} = \frac{S_{\hat{\beta}_i}}{N - 1} - \frac{1}{N}\sum_{i=1}^{N} \hat{\sigma}_i^2 (X_i'X_i)^{-1} \tag{5.11}$$

where

$$S_{\hat{\beta}_i} = \sum_{i=1}^{N} \hat{\beta}_i \hat{\beta}_i' - \frac{1}{N}\sum_{i=1}^{N} \hat{\beta}_i \sum_{i=1}^{N} \hat{\beta}_i' \tag{5.12}$$

Note that $S_{\hat{\beta}_i}/(N - 1)$ is the sample variance-covariance matrix of the β_i. To estimate Δ, the sample covariance matrix is adjusted by subtracting off some measure of variation due to the disturbances. This measure is given by the term

$$\frac{1}{N}\sum_{i=1}^{N} \hat{\sigma}_i^2 (X_i'X_i)^{-1} \tag{5.13}$$

Substituting Δ and the $\hat{\sigma}_i^2$ into equation (5.8) yields the feasible Aitken estimator of $\overline{\beta}$:

$$\hat{\overline{\beta}} = \left\{ \sum_{i=1}^{N} [\hat{\Delta} + \hat{\sigma}_i^2 (X_i'X_i)^{-1}]^{-1} \right\}^{-1} \sum_{i=1}^{N} [\hat{\Delta} + \hat{\sigma}_i^2 (X_i'X_i)^{-1}]^{-1} \hat{\beta}_i \tag{5.14}$$

The estimator $\hat{\Delta}$ in equation (5.11) is the difference between two matrices. Since Δ is a variance-covariance matrix it is assumed to be nonsingular with all its diagonal elements positive. The estimation procedure in equation (5.11) does not take these restrictions into account, however. It may occur that $\hat{\Delta}$ will yield negative estimates for variances of some of the coefficients.

Swamy (1971, pp. 107–110) suggests possible reasons why this may occur: 1) the assumed model may be incorrect or 2) statistical noise may be obscuring the underlying physical situation.

The assumed model may be incorrect in terms of the specification of the disturbances, ε_i. If disturbances are contemporaneously correlated or autocorrelated and these violations of the assumptions are not taken into account by the estimation procedure, negative variance estimates may result. Procedures to adjust for these violations will be discussed later.

An alternative possibility which could also be classified as a model misspecification is that certain coefficients may be random but others may be fixed and equal for all cross-sectional units. For those coefficients that are fixed, negative variance estimates may occur. Methods for treating a model with both fixed and random coefficients will be discussed in a later section.

An estimator of Δ could be constructed which includes the restriction that all diagonal elements must be nonnegative. However, it is unclear that the additional complexity of such an estimator would justify the gains in terms of the quality of the estimate of $\bar{\beta}$. Three alternatives to the restricted estimators are available:

1. Use the estimator

$$\hat{\Delta}_1 = \begin{cases} \hat{\Delta} & \text{if all diagonal elements are positive} \\ \hat{\Delta}_A & \text{with negative diagonal elements replaced by zero} \\ & \text{if negative variance estimates are present} \end{cases}$$

2. Use the estimator

$$\hat{\Delta}_2 = \frac{S_{\hat{\beta}_i}}{N - 1}$$

The estimator $\hat{\Delta}_2$ is simply $\hat{\Delta}$ without the subtraction of the term to adjust for variation arising from the disturbances.

3. Use the estimator

$$\hat{\Delta}_3 = \begin{cases} \hat{\Delta} & \text{if } \hat{\Delta} \text{ is positive definite} \\ \hat{\Delta} + (-\hat{\mu}_{min} + \eta)I & \text{otherwise} \end{cases}$$

where $\hat{\mu}_{min}$ is the smallest eigenvalue of $\hat{\Delta}$ and $\eta > 0$ is a small fixed number; see Havenner and Swamy (1981, p. 185)

Swamy has shown that the estimator, $\hat{\bar{\beta}}$, in equation (5.14) is consistent as both N and $T \to \infty$ and is asymptotically efficient as $T \to \infty$. Carter and Yang (1986) consider the case where the time series lengths may differ between individuals. Writing the number of time series observations for individual i as T_i, they show that equation (5.14) for computing $\hat{\bar{\beta}}$ still applies and that the estimator is consistent as N and $\min(T_i) \to \infty$ where $\min(T_i)$ represents the smallest time series length.

Carter and Yang (1986) also consider the case when the sample sizes T_i are small and N is large. If the assumption can be made that, $\sigma_i^2 = \sigma^2$ for $i = 1,\ldots,N$, that is, if the disturbance variances are equal, then the estimator $\hat{\bar{\beta}}$ is consistent as $N \to \infty$ and asymptotically efficient as $N \to \infty$. See also Johansen (1982).

Swamy (1974) examined the RCR model with alternative assumptions applied to the disturbances, ε_i. These assumptions replace Assumption 5.3 and may be stated:

Assumption 5.6: Assume that disturbances may be contemporaneously correlated as in the SUR model:

$$E(\varepsilon_i) = 0$$

$$E(\varepsilon_i \varepsilon_j) = \begin{cases} \sigma_i^2 I & \text{if } i = j \\ \sigma_{ij} I & \text{if } i \neq j \end{cases}$$

Assumption 5.7: Assume that disturbances may be autocorrelated:

$$\varepsilon_{it} = \rho_i \varepsilon_{i,t-1} + u_{it}$$

where ρ_i is the first-order autocorrelation coefficient.

With Assumptions 5.6 and 5.7 the variance-covariance matrix of the disturbances would be

$$E(\varepsilon\varepsilon') = \Omega = \begin{bmatrix} X_1'\Delta X_1 + \sigma_1^2\Omega_{11} & \sigma_{12}\Omega_{12} & \cdots & \sigma_{1N}\Omega_{1N} \\ \sigma_{21}\Omega_{21} & X_2'\Delta X_2 + \sigma_2^2\Omega_{22} & \cdots & \sigma_{2N}\Omega_{2N} \\ \vdots & \vdots & & \vdots \\ \sigma_{N1}\Omega_{N1} & \sigma_{N2}\Omega_{N2} & \cdots & X_N'\Delta X_N + \sigma_N^2\Omega_{NN} \end{bmatrix} \tag{5.15}$$

where

$$\Omega_{ij} = \frac{1}{1 - \rho_i \rho_j} \begin{bmatrix} 1 & \rho_i & \rho_i^2 & \cdots & \rho_i^{T-1} \\ \rho_j & 1 & \rho_i & \cdots & \rho_i^{T-2} \\ \vdots & \vdots & \vdots & & \vdots \\ \rho_j^{T-1} & \rho_j^{T-2} & \rho_j^{T-3} & \cdots & 1 \end{bmatrix} \tag{5.16}$$

Swamy derives the feasible Aitken estimator under Assumption 5.6 and 5.7 as

$$\hat{\bar{\beta}} = (X'\hat{\Omega}^{-1}X)^{-1}\, X'\hat{\Omega}^{-1}Y \tag{5.17}$$

where $\hat{\Omega}^{-1}$ is computed by replacing the unknown quantities in the Ω matrix by consistent estimators.

The matrix Δ can be estimated by

$$\hat{\Delta} = \frac{S_2}{N-1} - \frac{1}{N}\sum_{i=1}^{N} \hat{\sigma}_i^2 (X_i'\hat{\Omega}_{ii}^{-1}X_i)^{-1}$$

$$+ \frac{1}{N(N-1)} \sum_{\substack{i \neq j \\ i,j=1}}^{N} \hat{\sigma}_{ij} (X_i'\hat{\Omega}_{ii}^{-1}X_i)^{-1}\, X_i'\hat{\Omega}_{ii}^{-1}\hat{\Omega}_{ij}\hat{\Omega}_{jj}^{-1}X_j$$

$$\times (X_j'\hat{\Omega}_{jj}^{-1}X_j)^{-1} \tag{5.18}$$

where

$$S_2 = \sum_{i=1}^{N} b_i b_i' - \frac{1}{N}\sum_{i=1}^{N} b_i \sum_{i=1}^{N} b_i' \tag{5.19}$$

$$b_i = (X_i'\hat{\Omega}_{ii}^{-1}X_i)\, X_i'\hat{\Omega}_{ii}^{-1}\, Y_i \tag{5.20}$$

$\hat{\Omega}_{ij}$ is obtained from Ω_{ij} by replacing ρ_i by $\hat{\rho}_i$, where $\hat{\rho}_i$ is a consistent estimator of ρ_i (for example see equation (2.12));

$$\hat{\sigma}_{ij} = \frac{\hat{u}_i'\hat{u}_j}{T} \tag{5.21}$$

where

$$\hat{u}_i' = (\hat{u}_{i1}, \hat{u}_{i2}, \ldots, \hat{u}_{iT})',$$

$$\hat{u}_{i1} = (1 - \hat{\rho}^2)^{\frac{1}{2}} \hat{\varepsilon}_{i1}' \tag{5.22}$$

and

$$\hat{u}_{it} = \hat{\varepsilon}_{it} - \hat{\rho}_i \hat{\varepsilon}_{i,t-1} \quad \text{for} \quad t = 2, \ldots, T. \tag{5.23}$$

5.2 THE MIXED RANDOM COEFFICIENT REGRESSION MODEL

When regression coefficients vary between individuals, the RCR model allows the treatment of the coefficients as randomly distributed over cross-sectional units. The distribution of each coefficient can be described by estimating a few population parameters. If, however, the relationship between the dependent and any explanatory variable is fixed across all individuals, then the RCR model is no longer appropriate. Here the term "fixed" is used to mean that the coefficient values are equal for all individuals.

To allow for such a possibility, a model is needed which contains both random and fixed coefficients. This model will be referred to as the mixed RCR model.

The model in equation (5.1) is rewritten as

$$Y_i = X_{1i}\beta_{1i} + X_{2i}\beta_2 + \varepsilon_i \tag{5.24}$$

where

Y_i is a $T \times 1$ vector of observations on the dependent variable for the ith cross-sectional unit.

X_{1i} is a $T \times K$ matrix of observations on K_1 explanatory variables.

X_{2i} is a $T \times K_2$ matrix of observations on K_2 explanatory variables with $K_1 + K_2 = K$.

β_{1i} is a $K_1 \times 1$ vector of coefficients assumed to be random with mean $\bar{\beta}_1$ and variance-covariance matrix $\Delta_{\beta 1}$.

β_2 is a $K_2 \times 1$ vector of coefficients assumed to be fixed.
ε_i is a $T \times 1$ vector of random disturbances.

Assuming $\beta_{1i} = \bar{\beta}_1 + v_i$ the model in equation (5.24) can be rewritten as

$$Y_i = X_{1i}\bar{\beta}_1 + X_{2i}\beta_2 + e_i \tag{5.25}$$

$$= Z_i\bar{\gamma} + e_i \tag{5.26}$$

where

$$e_i = X_{1i}v_i + \varepsilon_i \tag{5.27}$$

$$Z_i = [X_{1i} \vdots X_{2i}] \tag{5.28}$$

and

$$\bar{\gamma} = \begin{bmatrix} \bar{\beta}_1 \\ \beta_2 \end{bmatrix} \tag{5.29}$$

The model in equation (5.26) applies to each of N cross-sections. These N individual equations can be combined as

$$Y = Z\bar{\gamma} + e \tag{5.30}$$

where

$$Y = \begin{bmatrix} Y_1 \\ Y_2 \\ \cdot \\ \cdot \\ \cdot \\ Y_N \end{bmatrix} \quad Z = \begin{bmatrix} Z_1 \\ Z_2 \\ \cdot \\ \cdot \\ \cdot \\ Z_N \end{bmatrix} = \begin{bmatrix} X_{11} & X_{21} \\ X_{12} & X_{22} \\ \cdot & \cdot \\ \cdot & \cdot \\ \cdot & \cdot \\ X_{1N} & X_{2N} \end{bmatrix} \quad \text{and}$$

$$e = \begin{bmatrix} e_1 \\ \cdot \\ \cdot \\ \cdot \\ e_N \end{bmatrix} \tag{5.31}$$

The variance-covariance matrix of e is

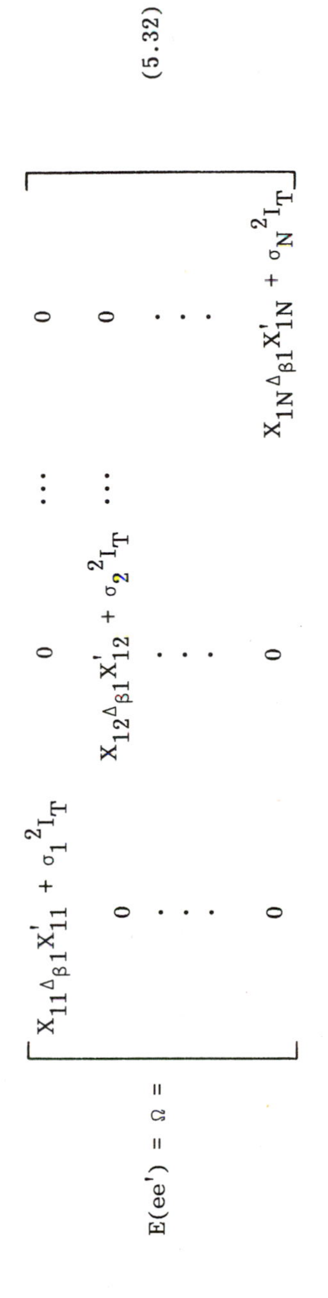

$$E(ee') = \Omega = \begin{bmatrix} X_{11}\Delta_{\beta 1}X'_{11} + \sigma_1^2 I_T & 0 & \cdots & 0 \\ 0 & X_{12}\Delta_{\beta 1}X'_{12} + \sigma_2^2 I_T & \cdots & 0 \\ \vdots & \vdots & & \vdots \\ 0 & 0 & & X_{1N}\Delta_{\beta 1}X'_{1N} + \sigma_N^2 I_T \end{bmatrix} \quad (5.32)$$

The GLS estimator of $\bar{\gamma}$ is

$$\tilde{\bar{\gamma}} = (Z'\Omega^{-1}Z)^{-1} Z'\Omega^{-1}Y \tag{5.33}$$

$$= \begin{bmatrix} X_1'\Omega^{-1}X_1 & X_1'\Omega^{-1}X_2 \\ X_2'\Omega^{-1}X_1 & X_2'\Omega^{-1}X_2 \end{bmatrix}^{-1} \begin{bmatrix} X_1'\Omega^{-1}Y \\ X_2'\Omega^{-1}Y \end{bmatrix}, \tag{5.34}$$

where

$$X_1 = \begin{bmatrix} X_{11} \\ X_{12} \\ \cdot \\ \cdot \\ \cdot \\ X_{1N} \end{bmatrix} \quad \text{and} \quad X_2 = \begin{bmatrix} X_{21} \\ X_{22} \\ \cdot \\ \cdot \\ \cdot \\ X_{2N} \end{bmatrix} \tag{5.35}$$

The mixed RCR model is, of course, simply a special case of the RCR model where the variances of certain coefficients are assumed to be equal to zero. Thus equation (5.8) still applies to estimation after certain elements af the Δ matrix are constrained to equal zero.

The matrix $\Delta_{\beta 1}$ can be estimated by

$$\hat{\Delta}_{\beta 1} = \frac{S_{\hat{\beta} 1}}{N - 1} - \frac{1}{N} \sum_{i=1}^{N} \hat{\sigma}_i^2 (X_{1i}'M_{2i}X_{1i})^{-1} \tag{5.36}$$

where

$$S_{\hat{\beta} 1} = \sum_{i=1}^{N} \hat{\beta}_{1i}\hat{\beta}_{1i}' - \frac{1}{N} \sum_{i=1}^{N} \hat{\beta}_{1i} \sum_{i=1}^{N} \hat{\beta}_{1i}' \tag{5.37}$$

$$\hat{\beta}_{1i} = (X_{1i}'M_{2i}X_{1i})^{-1}X_{1i}'M_{21}Y_i \tag{5.38}$$

and

$$M_{2i} = I_T - X_{2i}(X_{2i}'X_{2i})^{-1}X_{2i}' \tag{5.39}$$

Rosenberg (1973c) and Rayner and Wright (1980) examine mixed RCR models with the full rank assumption relaxed.

5.3 SMALL SAMPLE PROPERTIES OF THE RCR MODEL

The RCR and mixed RCR models have not been examined in as much detail as have the SUR and EC models. This is due to the additional complexity of the model. The exact finite sample distributions of the estimators are more difficult to derive and simulation studies require additional complexity in programming and more computer time.

Rao (1982) proposed conditions under which the estimator in equation (5.14) will be unbiased. In addition to assumptions 5.1 to 5.5 the following assumptions are added

Assumption 5.8: ε_i is a continuous random vector which follows a symmetric probability law.

Assumption 5.9: β_i is a continuous random vector which follows a symmetric probability law.

Rao shows $\hat{\bar{\beta}}$ will be unbiased provided assumptions 5.1 to 5.5, 5.8, and 5.9 hold and the expected value of $\hat{\bar{\beta}}$, $E(\hat{\bar{\beta}})$, exists. Rao gives conditions under which $E(\hat{\bar{\beta}})$ exists. To insure the existence of the expectation he recommends use of the estimator $\hat{\Delta}_3$ of Δ.

Johnson and Lyon (1973) used a Monte Carlo simulation to compare the performance of the classical pooling (CP) estimator, ANCOVA, estimation from data aggregated over cross-sectional units, error components (EC) and the RCR estimator. The models examined by Johnson and Lyon all include lagged values of the dependent variable as explanatory variables. That is, the basic model structure can be written as

$$y_{it} = \beta_0 + \beta_1 x_{it} + \beta_2 y_{i,t-1} + \varepsilon_{it} \tag{5.40}$$

for $i = 1,\ldots,N$ and $t = 1,\ldots,T$.

Swamy (1971, p. 143) noted that the RCR model was inappropriate in this case unless the regressors could be assumed "independent of their corresponding coefficients and the disturbance term. The condition that a regressor is independent of its own coefficient appears to be very strong and may be false in many occasions." When a dynamic relation such as that in equation (5.40) exists we can avoid this assumption by using the mixed RCR model where one assumes that the coefficients of the stochastic explanatory variable are fixed.

Johnson and Lyon do not make this assumption and go on to apply RCR estimation to their model. RCR estimation performed poorly overall in these cases as was anticipated by Swamy.

No other simulation experiments involving the RCR model are known to this author. To help provide some information as to the performance of the RCR model a simulation experiment was performed. The results of the simulation will be reported in Chapter 8.

5.4 INFERENCE IN THE RCR MODEL

The estimated variance-covariance matrix for the RCR model is

$$(X'\hat{\Omega}^{-1}X)^{-1} = \left\{ \sum_{i=1}^{N} [\hat{\Delta} + \hat{\sigma}_j^2 (X_i'X_i)^{-1}]^{-1} \right\}^{-1} \tag{5.41}$$

Standard errors for each coefficient can be obtained by taking the square root of the diagonal elements of the variance-covariance matrix. These can be used to construct tests on the means of coefficients. For example, to test whether the coefficients mean of the kth explanatory variable is equal to zero

$$H_0: \bar{\beta}_k = 0$$

versus

$$H_1: \bar{\beta}_k \neq 0 \tag{5.42}$$

the statistic used would be

$$t = \frac{\hat{\bar{\beta}}_k}{se(\hat{\bar{\beta}}_k)} \tag{5.43}$$

where $se(\hat{\bar{\beta}}_k)$ is the square root of the kth diagonal element of the variance-covariance matrix. Swamy (1971, pp. 120) has shown that the distribution of the statistic t is Student's t with N − 1 degrees of freedom. Swamy developed the test as a more general linear hypothesis test.

The hypothesis

$$H_0: \bar{\beta} = \bar{\beta}_0 \tag{5.44}$$

where $\bar{\beta}_0$ is some hypothesized value and $\bar{\beta}$ is the full vector of coefficient means was also developed by Swamy (1971, pp. 121-122). The test statistic used is

$$F = \frac{N - K}{K(N - 1)} (\hat{\bar{\beta}} - \bar{\beta}_0) \left\{ \sum_{i=1}^{N} [\hat{\Delta} + \hat{\sigma}_i^2 (X_i'X_i)^{-1}]^{-1} \right\} (\hat{\bar{\beta}} - \bar{\beta}_0) \tag{5.45}$$

Swamy showed that the asymptotic distribution (as $T \to \infty$) of the statistic is F with K and N − K degrees of freedom.

Carter and Yang (1986) further examined the properties of the test statistic in equation (5.45). They showed that the test statistic has the same asymptotic distribution when the time series lengths, T_i, differ between individuals, as long as $\min(T_i) \to \infty$. Also if the sample sizes T_i are small but $N \to \infty$, they suggest two possible test procedures. For a conservative test, use the test statistic in equation (5.45) as suggested previously. Alternatively, $[K(N - 1)/(N - K)]F$ has an approxi mate chi-square distribution with K degrees of freedom as $N \to \infty$ and can be used to test the hypothesis in equation (5.44). Since the use of the F-distribution to perform the test provides more conservative results, this is Carter and Yang's preferred procedure.

Swamy also examined tests for the randomness of coefficients. The hypothesis

$$H_0: \Delta = 0 \quad \text{given that} \quad E(\beta_i) = \bar{\beta} \quad \text{for all } i \tag{5.46}$$

is the hypothesis to be tested. Swamy develops the <u>likelihood ratio</u> (LR) criterion on which to base the test statistic. The LR criterion is

$$\Lambda = \frac{\sup_{H_0} L(\bar{\beta}, \Theta | Y, X)}{\sup_{\bar{\beta}, \Theta} L(\bar{\beta}, \Theta | Y, X)} \tag{5.47}$$

where Θ represents a vector of all unknown variances and covariances to be estimated and L is the likelihood function.

Under H_0, the estimates of $\bar{\beta}$ and the σ_i^2 are

$$\hat{\bar{\beta}}_0 = \left(\sum_{i=1}^{N} \frac{X_i'X_i}{\sigma_i^2} \right)^{-1} \left(\sum_{i=1}^{N} \frac{X_i'Y_i}{\sigma_i^2} \right) \tag{5.48}$$

and

$$\hat{\sigma}_{0i}^2 = \frac{(Y_i - X_i\hat{\bar{\beta}}_0)'(Y_i - X_i\hat{\bar{\beta}}_0)}{T} \tag{5.49}$$

where $\hat{\bar{\beta}}_0$ depends on the values of the $\hat{\sigma}_{0i}^2$ chosen. Swamy suggests starting with consistent estimators of the σ_i^2, say

$$\hat{\sigma}_i^2 = \frac{(Y_i - X_i\hat{\beta}_i)'(Y_i - X_i\hat{\beta}_i)}{T - K} \tag{5.50}$$

where $\hat{\beta}_i$ is the OLS estimator of β_i, and then solving equations (5.48) and (5.49) iteratively until stable estimates of $\bar{\beta}$ and the σ_i^2 are obtained. These can be substituted into the likelihood function to obtain

$$\sup_{\Delta=0} L(\bar{\beta},\Theta|Y,X) = (2\pi)^{-NT/2} \prod_{i=1}^{N} \sigma_i^{-T/2} \exp^{(-NT/2)} \tag{5.51}$$

The maximum likelihood estimates under the alternative hypothesis that $\Delta \neq 0$ are much more complex. For the nonlinear system of equations that must be solved to obtain the ML estimators the reader is referred to Swamy (1971, pp. 111-113). See Harville (1977) for a related discussion of ML estimation. The likelihood function in this case will be denoted

$$\sup_{\Delta\neq 0} L(\bar{\beta},\Theta|Y,X) \tag{5.52}$$

The likelihood ratio criterion is

$$\lambda = \frac{\sup_{\Delta=0} L(\bar{\beta},\Theta|Y,X)}{\sup_{\Delta\neq 0} L(\bar{\beta},\Theta|Y,X)}$$

Then, $-2 \ln \lambda$ will have a chi-square distribution with $K(K + 1)/2$ degrees of freedom. Because of the computational difficulties with the likelihood in equation (5.52), Swamy provides an alternate, asymptotically equivalent, form:

$$\lambda_A = \frac{\prod_{i=1}^{N} \hat{\sigma}_{0i}^{-T/2}}{\prod_{i=1}^{N} \hat{\sigma}_i^{-(T-K)/2} \, |X_i'X_i|^{-\frac{1}{2}} \left| \frac{S_{\hat{\beta}i}}{N} \right|^{-\frac{1}{2}}} \tag{5.53}$$

while $\hat{\sigma}_i$ is defined in equation (5.10) and $S_{\hat{\beta}i}$ is defined in equation (5.12).

Although the two criteria λ and λ_A are asymptotically equivalent, Dielman (1980, pp. 131–133) has provided some evidence that λ_A is not effective in small samples in choosing between models with fixed and random coefficients. An alternate indirect test proved more successful.

For the indirect test the null hypothesis to be tested is

$$H_0\colon \ \beta_1 = \beta_2 = \cdots = \beta_N = \beta \ \text{ given that } \Delta = 0 \tag{5.54}$$

The test procedure was first developed by Zellner (1962) for the SUR model. Zellner called it a test for aggregation bias. If the null hypothesis was accepted the data could be aggregated and the common coefficient vector estimated without any concern for aggregation bias.

In the RCR context, Swamy (1971, pp. 124–126) suggested that the test be used as a preliminary test to decide whether the assumptions of the RCR model may be reasonable. The test has also been used by Dielman (1980, Chapter 6), Dielman, Nantell, and Wright (1980) and Dielman and Oppenheimer (1984), among others, as a test for the randomness of coefficients. In such a case it must be assumed that, if the coefficients differ between individuals, then they can be modeled as random. If the coefficients are equal across all firms, then the RCR model would not be appropriate. The test is not directly applied to the Δ matrix so it was referred to as an indirect test for randomness.

The test statistic is

$$\sum_{i=1}^{N} \frac{(\hat{\beta}_i - \hat{\beta})' X_i' X_i (\hat{\beta}_i - \hat{\beta})}{\hat{\sigma}_i^2} \tag{5.55}$$

where $\hat{\beta}_i$ is the OLS estimate of β_i and

$$\hat{\beta} = \left(\sum_{i=1}^{N} \frac{X_i' X_i}{\hat{\sigma}_i^2} \right)^{-1} \sum_{i=1}^{N} \frac{X_i' Y_i}{\hat{\sigma}_i^2} \tag{5.56}$$

which is a form of the CP estimator. Thus $\hat{\beta}$ assumes coefficients are equal for all individuals and estimates the common coefficient vector β.

The asymptotic distribution of the statistic is well approximated by a Chi-square with $(N - 1)K$ degrees of freedom (for large values of T). Alternatively, Zellner (1962) has shown that if the statistic in equation (5.55) is divided by $(N - 1)K$, the resulting statistic can be used to test the null hypothesis in (5.54). This form of the statistic will have, asymptotically, an F-distribution with $(N - 1)K$ numerator and $N(T - K)$ denominator degrees of freedom. It is not known which form of the test will perform best in small samples.

Consider now the mixed RCR model of equation (5.24):

$$Y_i = X_{1i}\beta_{1i} + X_{2i}\beta_2 + \varepsilon_i \tag{5.57}$$

for $i = 1,\dots,N$. This model can be rewritten as

$$Y_i = Z_{1i}\gamma_{1i} + Z_{2i}\gamma_{2i} + X_{2i}\beta_2 + \varepsilon_i \tag{5.58}$$

where

$$\beta_{1i} = \begin{bmatrix} \gamma_{1i} \\ \gamma_{2i} \end{bmatrix} \qquad X_{1i} = \begin{bmatrix} Z_{1i} \\ Z_{2i} \end{bmatrix} \tag{5.59}$$

γ_{1i} is an $H_1 \times 1$ vector of random coefficients to be included in a test of some hypothesis; γ_{2i} is an $H_2 \times 1$ vector of random coefficients, but these are to be excluded from the test; Z_{1i} and Z_{2i} are, respectively, $T \times H_1$ and $T \times H_2$ matrices of observations on the independent variables; and all other terms were defined when discussing equation (5.24).

As previously noted, the mixed RCR model can be rewritten as

$$Y = Z_1\bar{\gamma}_1 + Z_2\bar{\gamma}_2 + X_2\beta_2 + e \tag{5.60}$$

where

$$Y = \begin{bmatrix} Y_1 \\ Y_2 \\ \cdot \\ \cdot \\ \cdot \\ Y_N \end{bmatrix} \quad Z_1 = \begin{bmatrix} Z_{11} \\ Z_{12} \\ \cdot \\ \cdot \\ \cdot \\ Z_{1N} \end{bmatrix} \quad Z_2 = \begin{bmatrix} Z_{21} \\ Z_{22} \\ \cdot \\ \cdot \\ \cdot \\ Z_{2N} \end{bmatrix} \quad X_2 = \begin{bmatrix} X_{21} \\ X_{22} \\ \cdot \\ \cdot \\ \cdot \\ X_{2N} \end{bmatrix}$$

$$e = \begin{bmatrix} e_1 \\ e_2 \\ \cdot \\ \cdot \\ \cdot \\ e_N \end{bmatrix} \tag{5.61}$$

and $\bar{\gamma}_1$ and $\bar{\gamma}_2$ are means of the random coefficients γ_{1i} and γ_{2i}. Note that the e_i consist of random components due to the disturbances ε_i as well as the random components of the coefficients, γ_{1i} and γ_{2i}, as in equation (5.27).

In the mixed RCR model, procedures are available to test the following hypothesis for randomness of coefficients:

$$H_0: \gamma_{11} = \gamma_{12} = \cdots = \gamma_{1N} \tag{5.62}$$

This is analogous to the indirect test for randomness in the RCR model. In this case, there may be a subset of coefficients which are initially assumed random but which are to be tested for randomness. If the null hypothesis is accepted we assume that the coefficients are fixed and should be treated in the manner of the β_2 vector of coefficients in equation (5.57). If the null hypothesis is rejected, then the coefficients γ_{1i} are treated as random.

The test statistic used to conduct the test is

$$\sum_{i=1}^{N} \frac{(\hat{\gamma}_{1i} - \hat{\gamma}_1)' Z_{1i}' Z_{1i} (\hat{\gamma}_{1i} - \hat{\gamma}_1)}{\sigma_i^2} \tag{5.63}$$

where $\hat{\gamma}_1$ is the estimated vector of coefficients assuming they are fixed and the $\hat{\gamma}_{1i}$ are the separate estimates of the coefficients.

Dielman (1980, pp. 35-39) has shown this statistic to have a limiting chi-square distribution with $(N - 1)H_1$ degrees of freedom.

To test a hypothesis on the means of a subset of coefficients

$$H_0: \bar{\gamma}_1 = \bar{\gamma}_{10} \tag{5.64}$$

where $\bar{\gamma}_{10}$ is a vector of preassigned values, the following statistic is used:

$$\frac{N - H_1}{H_1(N - 1)} (\hat{\bar{\gamma}}_1 - \bar{\gamma}_{10})' \hat{V}(\bar{\gamma}_1)^{-1} (\hat{\bar{\gamma}}_1 - \bar{\gamma}_{10}) \tag{5.65}$$

where $\hat{\bar{\gamma}}_1$ is the RCR estimate of $\bar{\gamma}_1$ and $V(\hat{\bar{\gamma}}_1)^{-1}$ is the inverse of the partition of the estimated variance-covariance matrix of the coefficients corresponding to the partition $\bar{\gamma}$.

Dielman (1980, pp. 31-35) has shown this statistic to have, asymptotically, an F-distribution with H_1 numerator and $N - H_1$ denominator degrees of freedom under the null hypothesis.

Although the above test has been framed as a test on the mean for a subset of random coefficients, the vector $\bar{\gamma}_1$ in (5.64) could just as well contain fixed coefficients. A test of whether these fixed coefficients are equal to some preassigned value can be performed simply by treating them as though they were means of random coefficients but with variances constrained to equal zero. The derivation of the test statistic and its distributional properties will remain the same.

See also Schmalensee (1972) for other hypothesis testing results.

5.5 OTHER RCR MODEL DEVELOPMENTS

In experimental settings the design matrices, X_i, are often the same for each individual. The RCR model has been examined in this case by Rao (1965), Norberg (1977), Johansen (1983), Fisk (1967) and Reinsel (1984, 1985).

Swamy (1973) discusses alternative estimators for the RCR model of Section 5.1. Bayesian estimation is discussed by Fearn (1975). Porter (1973) examines alternative estimators of the coefficient means when sampling designs other than simple random sampling are used.

Mundlak (1978b) considers the RCR model when explanatory variables and coefficients are correlated. He considers modeling the dependence through an auxiliary regression.

Laird and Ware (1982) examined the use of the EM algorithm to compute ML estimates in a model similar to the mixed RCR model discussed in this chapter.

Kadiyala and Oberhelman (1982) consider the problem of obtaining best linear unbiased (BLU) estimators of the individual coefficients, β_i. Using a theorem in Harville (1976) they show that the BLU estimator of β_i, assuming the variance-covariance matrix is known, is

$$\tilde{\beta}_i = \tilde{\bar{\beta}} + \Delta(\Delta + (X_i'X_i)^{-1})^{-1}\hat{\beta}_i - \Delta(\Delta + (X_i'X_i)^{-1})^{-1}\tilde{\bar{\beta}} \tag{5.66}$$

The RCR model specification in Section 5.1. is assumed. To operationalize the estimator, $\hat{\bar{\beta}}$ can be used in place of $\tilde{\bar{\beta}}$ and Δ can be replaced by an estimator, $\hat{\Delta}$:

$$\hat{\beta}_i^* = \hat{\bar{\beta}} + \hat{\Delta}(\hat{\Delta} + (X_i'X_i)^{-1})^{-1}\hat{\beta}_i - \hat{\Delta}(\hat{\Delta} + (X_i'X_i)^{-1})^{-1}\hat{\bar{\beta}} \tag{5.67}$$

The "*" is to distinguish this predictor from the OLS estimator $\hat{\beta}_i$. Note that $\hat{\beta}_i$ is the BLU estimator of β_i in Y_i. However, $\hat{\beta}_i^*$ is BLU in Y, the full vector of dependent variable values for all individuals. Kadiyala and Oberhelman use a Monte Carlo simulation to show that $\hat{\beta}_i^*$ is more efficient than $\hat{\beta}_i$ when N is large (N = 25 or 40). One can do just as well using $\hat{\beta}_i$ when N is small (N = 10 in the experiment).

Amemiya (1978) considers a case where the coefficients β_i are related to other explanatory variables through an equation such as

$$\beta_i = W_i\gamma_i + v_i \tag{5.68}$$

where W_i is a known matrix of variable values, γ_i is a vector of unknown parameters, and v_i is a vector of disturbances. See also Rayner and Wright (1980).

Liu and Tiao (1980) examine the autoregressive model of order one, (AR(1) model) written as

$$y_{it} = \beta_i y_{i,t-1} + \varepsilon_{it} \tag{5.69}$$

where $|\beta_i| < 1$ and the ε_{it}'s are assumed to be independent and identically distributed as $N(0, \sigma_\varepsilon^2)$. The coefficients, β_i, are assumed to be randomly distributed from a beta distribution. A Bayesian analysis is performed and generalizations to higher order models are discussed.

Li and Hui (1983) propose an alternative empirical Bayes procedure to estimate the AR(1) model in equation (5.69) or higher order generalizations. They use a simulation to show that both the Liu and Tiao estimator and the empirical Bayes estimator improve on least squares; the Liu and Tiao estimator appears superior to the empirical Bayes estimator. However, the Liu and Tiao estimator requires knowledge of the prior distribution of the β_i's whereas the empirical Bayes procedure does not.

5.6 APPLICATIONS

Several applications of Swamy's RCR and mixed RCR models have appeared in the finance literature. Studies of the effects of announcements on security return and risk have been conducted by Dielman, Nantell, and Wright (1980), Dielman and Oppenheimer (1984), and Oppenheimer and Dielman (1988). Also see the example in Section 5.7. Boness and Frankfurter (1977) used the RCR model to examine the concept of risk-classes in finance. Boot and Frankfurter (1972) used RCR to examine the optimal mix of short- and long-term debt for firms. Feige and Swamy (1974) applied the RCR model to estimate demand equations for liquid assets and Feige (1974) modeled the demand for demand deposits. Rayner (1981) proposed using RCR procedures to aid in predicting the systematic risk of securities.

Mehta, Narasimham, and Swamy (1978) and Kraft and Rodekohr (1978) study the regional demand for gasoline using the RCR model. Further comments on these articles can be found in Johnson (1980a), Greene (1980) and Kraft and Rodekohr (1980). Rodekohr (1979) also considers the demand for other fuels.

Ferguson and Leech (1978) used RCR to estimate a yield function for a pine tree plantation.

Schmalensee (1976) applied RCR in a study of expectation formation. Barth, Kraft, and Kraft (1979) use Swamy's RCR approach to estimate a price equation for the manufacturing sector of the United States. The cross-sectional units are manufacturing industries. Results are compared to estimates obtained through aggregation (see Chapter 1) and SUR (see Chapter 3). Hendricks, Koenker, and Poirier (1979) examine systematic variation in regression parameters relating the β_i to individual characteristics. The resulting model is similar to the RCR model.

Granger, Engle, Ramanathan, and Andersen (1979) and Ramanathan and Mitchem (1982) model hourly demand for electricity by individual households. Although the assumptions they make are those of the RCR model, they use least squares to estimate the model parameters due to the extremely large size of their sample. This results in estimates that are unbiased and consistent but not efficient.

Sheiner, Rosenberg, and Melmon (1972) use RCR procedures to estimate the parameters of a pharmacokinetic model used in establishing proper drug dosages for patients.

Johnson (1975) discusses the use of the RCR model to examine intraregional economic homogeneity.

See also Wittink (1977) and Berry and Trennepohl (1981).

The TSCS computer package discussed by Hall (1978) provides one procedure for estimating a random coefficient model.

Havenner and Herman (1977) discuss a routine to be used with the SPEAKEASY/FEDEASY processor (see Condie, 1977) for estimating a random coefficient model which also allows for first-order autocorrelation of the disturbances.

Johnson and Oakenfull (1978) also report availability of an RCR estimation program.

For other surveys which discuss RCR models see Johnson (1977, 1978, 1980b) and Spjotvoll (1977).

5.7 EXAMPLE: TENDER OFFER MERGERS AND STOCKHOLDER'S WEALTH

Most of the abundant finance literature on mergers can be classified into two categories: 1) research on the motivation for mergers, and 2) research on the impact of mergers on the wealth of the stockholders. Without passing judgment on any hypotheses as to why firms merge, this study uses the RCR methodology to take a new look at the question of mergers and stockholder wealth. The RCR approach provides some new insights into the conflicting empirical results which have previously been published.

For a sample on nonconglomerate mergers, Halpern (1973), through the use of fairly standard "residual analysis," concludes that there are abnormal gains to the stockholders of both firms involved in a merger and that these wealth benefits are approximately evenly distributed. On the other hand, Mandelker (1974) reports that only the acquired firms in his sample earned abnormal returns around the time of the merger announcement. Dodd and Ruback (1977), using a sample of mergers accomplished via tender offers, conclude that both the bidding firm and the target firm generate abnormal returns for their stockholders around the announcement of the merger. However, comparing abnormal returns for acquiring and acquired firms to those for control samples, Langetieg (1978) finds that for his sample, neither firms' stockholders earn significantly abnormal returns. As a final blow to an effort to find some consistency among all this research, Dodd (1980), using the announcement of a merger discussion as the announcement date, finds that the acquiring firm's stockholders actually earn slightly negative returns while the acquired firm's stockholders earn abnormal returns.

In a slightly different line of research, Gahlon and Stover (1979) report that the systematic risk for a sample of conglomerates did not change as a result of their merger activity, while Langetieg, Haugen, and Wichern (1980) report that the acquiring firms in their sample experienced an increase in their systematic risk.

Why are the empirical results so ambiguous? Reviewing the other major category of research on merger activity provides some interesting insights. Motivations for mergers are almost as numerous as mergers themselves. In general, the literature seems to explain merger activity in terms of management utility maximization and/or stockholder wealth maximization. However, such a dichotomy begs the more fundamental question as to exactly why merger activity should increase manager utility or stockholder wealth. On the stockholder wealth side, one hypothesis is that mergers make it easier for firms to gain monopolistic power in the market for its products. A second hypothesis is that firms merge in order to obtain real economies of scale in the production, distribution, and/or marketing of its products. A third hypothesis is that there have been times when the major driving force for merger activity had to do with the purchase of tax shields. Yet another hypothesis is that merging provides access to real risk reduction not obtainable by investors directly. Finally, mergers are seen by some to be a means of increasing the efficiency with which some firms are run. The efficiency gains occur when "bad" management is replaced and assets and/or financing are managed more productively by the new managers.

All of the above hypotheses are plausible, and, individually or jointly, in an efficient capital market they could result in various amounts of abnormal gains for the merging firms. How those gains would be distributed between the "buyer" and the "seller" depends on the magnitude of the gain and on the price paid for acquisition. If a premium above market price is paid for the acquired firms, its stockholders will enjoy abnormal gains. However, the acquiring firm's stockholders will experience zero, positive, or negative gains depending on the relationship between the economic gains derived from the various sources mentioned above and the price paid to achieve them.

In addition to the ambiguity caused by the relationship between benefits and prices in an efficient market, further ambiguity is suggested by mergers motivated by an attempt to profit from an inefficient capital market. The most commonly mentioned hypotheses here relate to accountants using merger activity to manipulate EPS numbers, and to a desire to take advantage of the market's misunderstanding of P/E ratios by having a high P/E company buy a low P/E company. If markets are in fact efficient, mergers motivated by these factors will almost certainly result in negative price performance for the buyer. The best that could be achieved would be a neutral influence. On the other hand, if the market is inefficient in this regard during at least some of the time periods studied, positive abnormal price changes could be observed.

As a final explanation for the ambiguity in the empirical results, we return to the maximization of management utility as a possible driving force for mergers. Attempts to diversify (in a manner that stockholders could achieve themselves) and attempts to expand the size of organizations could result in lowering the risk managers face and increasing their salary compensation even though no benefits accrue to stockholders. In such mergers, positive, zero, or negative price performance could be observed depending on how efficient the capital market's reaction is and depending on the price paid for the acquisition.

What is striking about all this is that all of these hypotheses have been discussed in the empirical literature reviewed above, all of these hypotheses are at least plausible, and yet, all the empirical work is performed as though each sample of firms is homogenous in the sense that the mergers were all motivated by the same factor and that this single motivating force can be identified by examining the mean price reaction of the sample. It is of course not only possible, but likely that motivations for merging will differ for different firms in any sample. To draw conclusions from the statistics derived from using a heterogeneous sample ignores the problem that significant price effects for different firms with different motivations may, for example, simply be offsetting one another. In such a situation, it might be concluded that the price effects of a merger are neutral, when in fact they are positive for some firms in the sample and negative for others.

What is needed is a methodology allowing investigation of whether the price effects of a merger are fixed or whether they vary from firm to firm, and, in either case, to determine the magnitude and sign of the effects. The methodology to be used here is that of RCR analysis, which specifically accounts for variation in price reaction between securities.

The Model The basic model to be estimated is

$$\begin{aligned} R_{i,t} - R_{F,t} &= a_i + b_i(R_{m,t} - R_{F,t}) \\ &\quad + m_i M_{i,t} \\ &\quad + p_{1;i} P_{1;i,t} + \cdots + p_{n;i} P_{n;i,t} \\ &\quad + s_{1;i} S_{1;i,t} + \cdots + s_{n;i} S_{n;i,t} + e_{i,t} \end{aligned} \tag{5.70}$$

where $M_{i,t}$ is equal to one if firm i announces a merger in month t, and is equal to zero for all other months, $P_{k;i,t}$ is equal to one if firm i announces a merger in month t + k, and equal to zero for all other months, and $S_{k;i,t}$ is equal to one if firm i announced a merger in month t − k, and is equal to zero for all other months.

In this model, the coefficient m_i of $M_{i,t}$ measures any excess return in the month of announcement, that is, any immediate effect. Dummy variables have been added to check on returns subsequent to the announcement. For instance, the explanatory variable $S_{1;i,t}$ is a dummy variable indicating the first month subsequent to announcement. If its coefficient, $s_{1;i}$, equals zero, then it can be concluded that any excess return during the announcement month was not canceled during the month following the announcement. In addition, a series of dummy variables, $P_{k;i,t}$, was included to test whether there were any information leakages prior to the announcement.

This model will be estimated for the pooled cross-sectional time series sample of monthly returns on firms that merged. The importance of the RCR model is that each regression coefficient is allowed to vary between firms and is viewed as a random variable. As is usually the case in the market model, the parameters a_i and b_i are allowed to vary across firms. The characteristic of this model of most interest though, is that the coefficients of the variables related to merger activity are also allowed to vary across firms.

As in random effects analysis of variance models, negative variance estimates may be obtained. Those coefficients with negative variance estimates are treated as fixed coefficients; their variances are constrained to be zero. A new model is then re-estimated containing both fixed and random coefficients. Results are reported only after appropriate zero variance constraints have been imposed.

Shifting Beta One of the more important methodological problems addressed in some of the previous research on the wealth effects of mergers is that of adjusting for changes in the level of risk that occur specifically because of the merger activity. A shift in risk could occur for a number of reasons. Most directly, combining two separate firms with their own risk characteristics into a single two-security portfolio must, by necessity, result in a new risk level which is some weighted average of the original risk levels. (This is referred to as the "weighting effect" on risk.) In addition, it has been suggested that use of increased leverage in order to finance the merger would result in an increased risk level. (This is referred to as the "financing effect" on risk.) Finally, risk may be reduced in some real way through improved management of assets. (This is referred to as the "real effect" on risk.)

Whatever the direction of the change in risk and whatever its cause, empirical attempts to identify risk-adjusted wealth effects of mergers must account for any shifts in risk that occur. Due to the use of RCR methodology a straightforward adjustment to the model allows direct adjustment for shifts in beta, the risk measure. A single term is added to the model as given in equation (5.70), $c_i C_{i,t}$ where

$$C_{i,t} = \begin{cases} R_{m,t} - R_{f,t} & \text{for time periods including and following the period of merger announcement} \\ 0 & \text{for the time periods before the merger} \end{cases}$$

The coefficient of $C_{i,t}$, c_i, represents any change in beta which occurs at or near the announcement period and is maintained thereafter.

To better illustrate the use of the varialbe $C_{i,t}$, the beta coefficient in these models can be viewed as having been decomposed into two parts:

$$\text{"beta"} = \begin{cases} b_i & \text{before the merger announcement} \\ b_i + c_i & \text{for the time periods of and subsequent to the announcement} \end{cases}$$

The "beta" of each firm is therefore the sum of the effects accounted for by $R_{m,t} - R_{F,t}$.

Data The period from 1960 through 1978 was scanned to find firms that merged by way of tender offers. The source of information was a list compiled annually by the NYSE which identifies all NYSE firms that were delisted due to consolidation with other NYSE firms. Given this list of mergers, the Wall Street Journal was searched to identify the date of the merger announcement. From this process 83 mergers by tender offer were identified.

For each of the firms, monthly returns were collected for eight years prior to each firm's first merger, and for up to five years subsequent to each firm's final merger. These returns included dividends as well as capital gains. Market returns were taken from the CRSP data set. Risk-free rates ($R_{F,t}$) are monthly rates on three-month treasury bills. The risk-free rate is needed because the models are actually fit to excess returns (above risk-free level) on the securities and on the market.

Results For the surviving firms, mergers consummated by tender offers appear to have no significant impact on the stockholders (see Table 5.1). However, the sample is not a homogeneous sample. In other words, the results do vary cross-sectionally as indicated by the positive population standard deviation for m_i. Further research is required in order to explain the randomness of this coefficient before it can be safely concluded that surviving firms in general do not benefit from mergers by way of tender offers. Mergers with some set of characteristics may benefit surviving firm stockholders while mergers with other characteristics result in lower returns. At any rate, the merger effects are not fixed, although on average they do seem to equal zero. Finally, it should be emphasized that if the significant increase in risk, as indicated by the value of the c_i coefficient for the survivor sample, had not been accounted for, the announcement effect would have been mistakenly estimated at a higher and perhaps significant level.

The results for the acquired firm's stockholders show large positive benefits in the month of and in the month preceeding the merger announcement, and the results vary significantly across the sample. Once again, it cannot be concluded in general that such mergers benefit selling stockholders, only that they do on average. There may be some type of merger that does not benefit the stockholders. To investigate this further the sample was segmented into a group of firms with only one potential acquirer and

TABLE 5.1 RCR Estimates for Tender Offers

Acquired Sample

Coefficient	Mean	Population Standard Deviation[a]	Test for Mean Zero, t
a_i	0.002	*	2.34
b_i	1.085	0.299	27.53
c_i	-0.136	0.423	-1.67
m_i	0.158	0.144	8.79
$p_{1,i}$	0.054	0.057	5.30

Survivor Sample

Coefficient	Mean	Population Standard Deviation[a]	Tet for Mean Zero, t
a_i	0.002	*	3.79
b_i	1.152	0.309	27.28
c_i	0.155	0.284	3.18
m_i	0.011	0.045	1.09
1% Critical Value			2.6

[a]Coefficients constrained to be fixed are indicated by an asterisk, *.

a group of firms with tender offers from more than one firm. There were 43 firms in the former group and 40 in the latter. The results of the analysis for these two groups are shown in Table 5.2. As can be seen this segmentation of the sample doesn't help much in reducing the variability of the announcement month effect (although the estimated population standard deviation for the single offer firms is somewhat smaller).

TABLE 5.2 RCR Estimates for Fragmented Sample

Single Offer Acquired Firms

Coefficient	Mean	Population Standard Deviation[a]	Test for Mean Zero, t
a_i	0.002	*	1.84
b_i	1.141	0.385	16.91
m_i	0.131	0.116	6.06
$p_{1,i}$	0.046	0.064	3.00

Multiple Offer Acquired Firms

Coefficient	Mean	Population Standard Deviation[a]	Test for Mean Zero, t
a_i	0.0002	*	0.24
b_i	1.034	0.164	27.27
m_i	0.188	0.168	6.50
$p_{1,i}$	0.064	0.049	4.94
1% Critical Value			2.7

[a]Coefficients constrained to be fixed are indicated by an asterisk, *.

For the single offer firms the average increase in return in the announcement month is 13.1% while the average increase for multiple offer firms is 18.8%. A t-test was used to compare these two averages. The t-statistic value was 1.59 (with p-value 0.1118).

Summary A random coefficient regression model was used to investigate the effects of tender offer mergers on the rate of return to the stockholders of the acquired firm and of the surviving firm firm. The results are:

1. Tender offer mergers have no significant effect on returns to stockholders of the surviving firm; however, there is significant variation in this effect across mergers. In other words, although the average result is neutral, there may be mergers motivated by certain characteristics which result in positive return effects and others which result in negative return effects.

2. Tender offer mergers have a significant positive effect on the returns to stockholders of the acquired firm; however, there is a significant variation in the effect across mergers. Again, such a result suggests a search for those characteristics which will reduce the variation in announcement effects.

3. The risk of the surviving firms increased for mergers consummated by tender offers.

4. Segmenting the sample into those firms with only one tender offer and those with multiple tender offers was not very helpful in reducing variation in the announcement effects.

5. Although firms with multiple tender offers had a greater month of announcement return effect, on average, there is not sufficient evidence (for a level of significance less than about 11%) to conclude that the average effect is significantly different from that of the single tender offer firms.

These results begin to explain the ambiguous nature of the results presented in some of the previous literature on mergers. For tender offer mergers, the motivations are apparently too diverse to allow researchers to treat them as a homogeneous sample. The return effects of merger announcements were not fixed, as is usually assumed, and varied significantly across mergers. An explanation of this variation is the subject of continuing research.

6
Time and Cross-Sectionally Varying Parameter Models

6.1 INTRODUCTION

Various authors have extended the idea of coefficient variation to the time dimension. Swamy and Mehta (1977a) motivate time and cross-sectionally varying parameter models as follows:

> We allow for different coefficients for each individual unit to account for spatial or interindividual heterogeneity, and further, we modify continually the values of coefficients over time so as to allow the relationship to adapt itself to local conditions.

The time and cross-sectionally varying parameter models are the most general of the models for analyzing pooled data. They are also the most difficult to handle notationally, computationally, and analytically. In this chapter, models with time and cross-sectionally varying parameters are examined under the assumptions proposed by several authors.

6.2 THE HSIAO MODEL

6.2.1 The Model

Hsiao (1972, 1974, 1975) proposed a model that views each regression coefficient realization as a random draw from a population with mean $\bar{\beta}$ and with random components v_i and u_t. The model is written as

$$y_{it} = x_{it}\beta_{it} + \varepsilon_{it} \tag{6.1}$$

where

y_{it} is the tth observation of the dependent variable for the ith cross-sectional unit.

x_{it} is a $1 \times K$ vector of observations on the independent variables.

β_{it} is a $K \times 1$ vector of regression coefficients.

ε_{it} is a random disturbance.

The regression coefficients, β_{it}, can be written as

$$\beta_{it} = \bar{\beta} + v_i + u_t \tag{6.2}$$

where

$$\bar{\beta} = (\bar{\beta}_1, \bar{\beta}_2, \ldots, \bar{\beta}_K)' \tag{6.3}$$

is the common mean of the coefficient vectors,

$$v_i = (v_{1i}, v_{2i}, \ldots, v_{Ki})' \tag{6.4}$$

is constant through time but differs among individuals, and

$$u_t = (u_{1t}, u_{2t}, \ldots, u_{Kt})' \tag{6.5}$$

differs over time but remains constant among individuals.

Hsiao's assumptions for the model in (6.1) and (6.2) are as follows:

Assumption 6.1: $E(\varepsilon_{it}) = 0$

Assumption 6.2: $E(\varepsilon_{it}\,\varepsilon_{js}) = \begin{cases} \sigma^2 & \text{if } i = j,\ t = s \\ 0 & \text{otherwise} \end{cases}$

Assumption 6.3: $E(v_i) = E(u_t) = 0$

Assumption 6.4: $E(v_i v_j) = \begin{cases} \Delta_v & \text{if } i = j \\ 0 & \text{otherwise} \end{cases}$

where Δ_v is a diagonal matrix

Assumption 6.5: $E(u_t\ u_s) = \begin{cases} \Delta_u & \text{if } t = s \\ 0 & \text{otherwise} \end{cases}$

where Δ_u is a diagonal matrix

Assumption 6.6: The v_i, u_t, and ε_{it} are independent.

Define the following matrices and vectors:

$$Y = \begin{bmatrix} Y_1 \\ Y_2 \\ \cdot \\ \cdot \\ \cdot \\ Y_N \end{bmatrix} \quad \text{where } Y_i = \begin{bmatrix} y_{i1} \\ y_{i2} \\ \cdot \\ \cdot \\ \cdot \\ y_{iT} \end{bmatrix} \tag{6.6}$$

$$X = \begin{bmatrix} X_1 \\ X_2 \\ \cdot \\ \cdot \\ \cdot \\ X_N \end{bmatrix} = \begin{bmatrix} x_{111} & \cdots & x_{1K1} \\ x_{112} & \cdots & x_{1K2} \\ \vdots & \ddots & \vdots \\ x_{11T} & \cdots & x_{1KT} \\ \vdots & & \vdots \\ x_{N11} & \cdots & x_{NK1} \\ x_{N12} & \cdots & x_{NK2} \\ \vdots & \ddots & \vdots \\ x_{NIT} & \cdots & x_{NKT} \end{bmatrix}_{NT \times K} \tag{6.7}$$

$$\underline{X}_1 = \begin{bmatrix} X_1 & 0 & \cdots & 0 \\ 0 & X_2 & \cdots & 0 \\ \vdots & \vdots & & \vdots \\ 0 & 0 & \cdots & X_N \end{bmatrix}_{NT \times NK} \tag{6.8}$$

$$\underline{X}_2 = \begin{bmatrix} X_{1\cdot 1} & 0 & \cdots & 0 \\ 0 & X_{1\cdot 2} & \cdots & 0 \\ \vdots & \vdots & \ddots & \vdots \\ 0 & 0 & \cdots & X_{1\cdot T} \\ X_{2\cdot 1} & 0 & \cdots & 0 \\ 0 & X_{2\cdot 2} & \cdots & 0 \\ \vdots & \vdots & \ddots & \vdots \\ 0 & 0 & \cdots & X_{2\cdot T} \\ \vdots & & & \vdots \\ X_{N\cdot 1} & 0 & \cdots & 0 \\ 0 & X_{N\cdot 2} & \cdots & 0 \\ \vdots & \vdots & \ddots & \vdots \\ 0 & 0 & \cdots & X_{N\cdot T} \end{bmatrix}_{NT \times TK} \tag{6.9}$$

where

$$X_{i\cdot t} = [x_{i1t} \; x_{i2t} \; \cdots \; x_{iKt}] \tag{6.10}$$

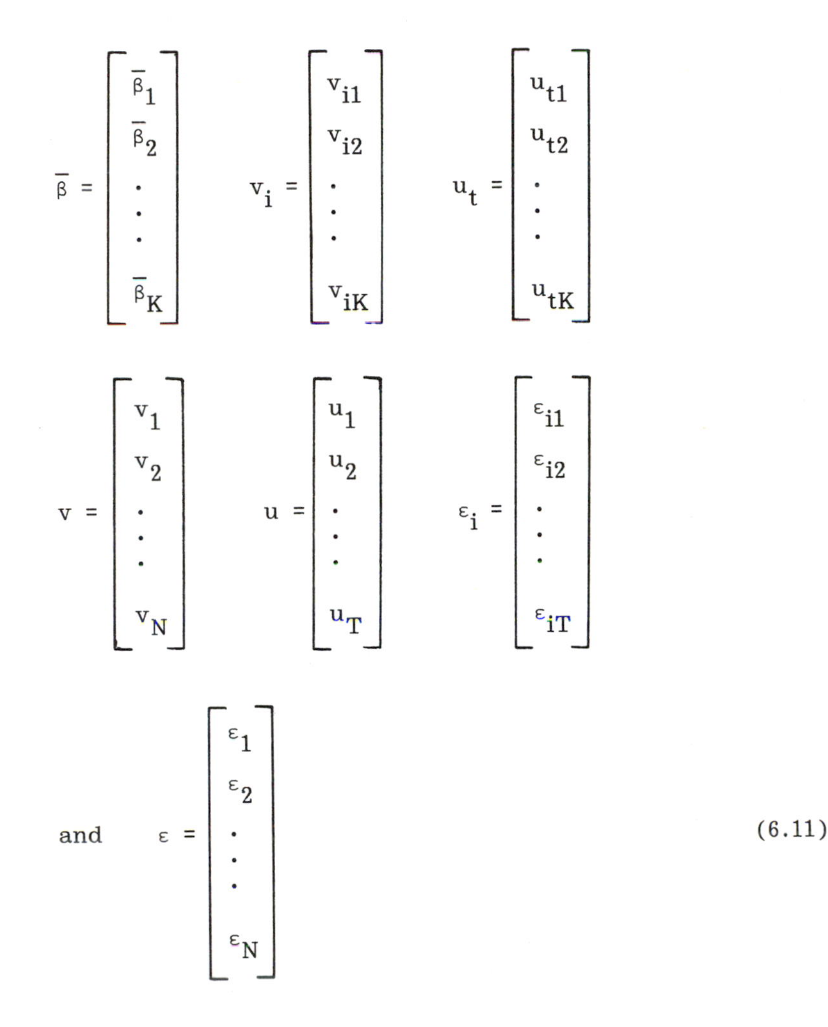

$$\bar{\beta} = \begin{bmatrix} \bar{\beta}_1 \\ \bar{\beta}_2 \\ \cdot \\ \cdot \\ \cdot \\ \bar{\beta}_K \end{bmatrix} \qquad v_i = \begin{bmatrix} v_{i1} \\ v_{i2} \\ \cdot \\ \cdot \\ \cdot \\ v_{iK} \end{bmatrix} \qquad u_t = \begin{bmatrix} u_{t1} \\ u_{t2} \\ \cdot \\ \cdot \\ \cdot \\ u_{tK} \end{bmatrix}$$

$$v = \begin{bmatrix} v_1 \\ v_2 \\ \cdot \\ \cdot \\ \cdot \\ v_N \end{bmatrix} \qquad u = \begin{bmatrix} u_1 \\ u_2 \\ \cdot \\ \cdot \\ \cdot \\ u_T \end{bmatrix} \qquad \varepsilon_i = \begin{bmatrix} \varepsilon_{i1} \\ \varepsilon_{i2} \\ \cdot \\ \cdot \\ \cdot \\ \varepsilon_{iT} \end{bmatrix}$$

and $$\varepsilon = \begin{bmatrix} \varepsilon_1 \\ \varepsilon_2 \\ \cdot \\ \cdot \\ \cdot \\ \varepsilon_N \end{bmatrix} \tag{6.11}$$

Hsiao's model can now be written as

$$Y = X\bar{\beta} + \underline{X}_1 u + \underline{X}_2 v + \varepsilon = X\bar{\beta} + e \tag{6.12}$$

where

$$e = \underline{X}_1 u + \underline{X}_2 v + \varepsilon.$$

The variance-covariance matrix can be written as

$$E(ee') = \Omega = \begin{bmatrix} X_1\Delta_v X_1' & 0 & \cdots & 0 \\ 0 & X_2\Delta_v X_2' & \cdots & 0 \\ \vdots & \vdots & \ddots & \vdots \\ 0 & 0 & \cdots & X_N\Delta_v X_N' \end{bmatrix}$$

$$+ \begin{bmatrix} D(X_1\Delta_u X_1') & D(X_1\Delta_u X_2') & \cdots & D(X_1\Delta_u X_N') \\ D(X_2\Delta_u X_1') & D(X_2\Delta_u X_2') & \cdots & D(X_2\Delta_u X_N') \\ \vdots & \vdots & \ddots & \vdots \\ D(X_N\Delta_u X_1') & D(X_N\Delta_u X_2') & \cdots & D(X_N\Delta_u X_N') \end{bmatrix} + \sigma^2 I_{NT} \tag{6.13}$$

where

$$D(X_i\Delta_u X_j') = \begin{bmatrix} X_{i\cdot 1}\Delta_u X_{j\cdot 1}' & 0 & \cdots & 0 \\ 0 & X_{i\cdot 2}\Delta_u X_{j\cdot 2}' & \cdots & 0 \\ \vdots & \vdots & \ddots & \vdots \\ 0 & 0 & \cdots & X_{i\cdot T}\Delta_u X_{j\cdot T}' \end{bmatrix} \tag{6.14}$$

The generalized least squares estimator of $\bar{\beta}$ in (6.12) is

$$\tilde{\bar{\beta}} = (X'\Omega^{-1}X)^{-1}\, X'\Omega^{-1}Y \tag{6.15}$$

In practice Ω must be estimated. Hsiao examines two estimation methods for the variance and covariance components in the model: the minimum norm quadratic unbiased estimation (MINQUE) method of Rao (1971a and 1972) and the maximum likelihood (ML) method.

See Kelejian and Stephan (1983) for further results on the Hsiao model.

6.2.2 MINQUE Estimation of Variance and Covariance Components

Hsiao's MINQUE estimator of the varaince and covariance components in Ω is essentially a Hildreth and Houck (1968) type estimator.

Consider first the matrix Δ_u. We can write the equation for the ith cross-sectional unit as

$$y_{it} = \sum_{k=1}^{K} (\bar{\beta}_k + v_{ik} + u_{kt}) x_{ikt} + \varepsilon_{it} \tag{6.16}$$

$$= \sum_{k=1}^{K} \beta_{ik} x_{ikt} + \eta_{it} \tag{6.17}$$

where

$$\beta_{ik} = \bar{\beta}_k + v_{ik} \tag{6.18}$$

and

$$\eta_{it} = \sum_{k=1}^{K} u_{kt} x_{itk} + \varepsilon_{it} \tag{6.19}$$

Also write

$$\beta_{i\cdot} = (\beta_{i1}, \ldots, \beta_{iK})' \qquad \text{for } i = 1, \ldots, N. \tag{6.20}$$

Let Θ_i be the T × T covariance matrix for η_{it}. Hsiao notes that the s − tth element of Θ_i will be

$$\Theta_{ist} = 0 \qquad \text{for } s \neq t \tag{6.21}$$

$$\Theta_{itt} = \sum_{k=1}^{K} x_{ikt}^2 \ \delta_{ukk} + \sigma^2 \tag{6.22}$$

$$= \dot{Z}_{(it)} \ \delta_u \tag{6.23}$$

where $\dot{Z}_i$ is a T × (K + 1) matrix whose elements are squares of the corresponding elements of $Z_i = (X_i, 1)$, $\dot{Z}_{(it)}$ is the tth row of Z_i and δ_u is the (K + 1) × 1 column vector with elements $\delta_{u11}, \ldots, \delta_{uKK}, \sigma^2$. Note also that $\delta_{u11}, \ldots, \delta_{uKK}$ are the diagonal elements of Δ_u, that is,

$$\Delta_u = \begin{bmatrix} \delta_{u11} & 0 & \cdots & 0 \\ 0 & \delta_{u22} & \cdots & 0 \\ \vdots & \vdots & \ddots & \vdots \\ 0 & 0 & \cdots & \delta_{uKK} \end{bmatrix} \tag{6.24}$$

Hsiao suggests estimating δ_u by first running an OLS regression for each cross-sectional unit over time, obtaining estimates of the β_{ik}'s and then computing the residuals as

$$\hat{\eta}_i = Y_i - Z_i\hat{\beta}_{i\cdot} = M_iU_i \tag{6.25}$$

where

$$M_i = I_T - Z_i(Z_i'Z_i)^{-1}Z_i \tag{6.26}$$

$$U_i = (\eta_{i1}, \ldots, \eta_{iT})' \tag{6.27}$$

Second, square each element of the vector of residuals, $\hat{\eta}_i$, and denote the vector of squared residuals as $\hat{\eta}_i^2$.

Either one of the following two estimators will provide unbiased estimators of δ_u:

$$\hat{\delta}_{ui} = (G_i' G_i)^{-1} G_i' \hat{\eta}_i^2 \tag{6.28}$$

where

$$G_i = M_i^2 Z_i \tag{6.29}$$

and M_i^2 is the matrix obtained by squaring each element of M_i.

$$\hat{\delta}_{ui} = (Z_i' M_i Z_i)^{-1} Z_i' \hat{\eta}_i^2 \tag{6.30}$$

Third, either average the $\hat{\delta}_{ui}$ obtained from one of the above two methods to obtain a consistent estimator of δ_u:

$$\hat{\delta}_u = \frac{1}{N} \sum_{i=1}^{N} \hat{\delta}_{ui} \tag{6.31}$$

or regress $\hat{\eta}^2 = (\hat{\eta}_1{}^2, \ldots, \hat{\eta}_N{}^2)'$ on $G = (G_1, \ldots, G_N)'$ to obtain

$$\hat{\delta}_u = (G'G)^{-1} \, G'\hat{\eta}^2 \tag{6.32}$$

Both methods result in consistent estimators of $\delta_u = (\delta_{u11}, \ldots, \delta_{uKK}, \sigma^2)$.

To estimate the elements of Δ_v, Hsiao recommends the same approach as above. Simply apply the same methods to equation (6.16) across individuals for a given time period.

Rewriting equation (6.16) as

$$y_{it} = \sum_{k=1}^{K} \beta_{kt} \, x_{itk} + \gamma_{it} \tag{6.33}$$

where

$$\beta_{kt} = \bar{\beta}_k + u_{kt} \tag{6.34}$$

and

$$\gamma_{it} = \sum_{k=1}^{K} v_{ik} \, x_{itk} + \varepsilon_{it} \tag{6.35}$$

it can be seen that the problem of estimating $\delta_v = (\delta_{v11}, \ldots, \delta_{vkk}, \sigma^2)'$ is symmetric to the previous problem.

Once δ_u and δ_v have been estimated, two estimates of σ^2 will be available. Hsiao suggests using the simple average of these two estimates.

6.2.3 ML Estimation of Variance and Covariance Components

Hsiao derives the ML estimator of the variance and covariance components assuming that all random components are normally distributed.

The likelihood function of Y in equation (6.6) can be written as

$$(2\pi)^{-(1/2)NT} |\Omega|^{-1/2} \exp[-\frac{1}{2}(Y - X\bar{\beta})'\Omega^{-1}(Y - X\bar{\beta})] \tag{6.36}$$

The log likelihood function can be shown to be proportional to

$$-NT \log 2\pi - \log |\Omega| - \text{trace } \Omega^{-1} (Y - Xb)(Y - Xb)' \tag{6.37}$$

where b is some estimator of $\bar{\beta}$ and it is assumed that $b = \bar{\beta}$. For example, b could be the OLS estimator of $\bar{\beta}$.

Maximizing the likelihood in equation (6.36) with respect to the variance and covariance components is achieved by differentiating the log likelihood in (6.37) with respect to each of the components, setting the resulting equations equal to zero and solving them simultaneously. The resulting equations will be nonlinear, obviously, and difficult to solve. Hsiao recommends an iterative procedure suggested by Anderson (1971) to obtain the solutions.

6.2.4 Inference Using Hsiao Model

Hsiao developed tests of certain hypotheses for his model based on both MINQUE and ML estimation.

One question to be asked concerning the model specification is whether the coefficients are random over time. The hypothesis to be tested would be

$$H_0: \Delta_u = 0 \tag{6.38}$$

Using MINQUE estimation, Hsiao proposes an indirect test of the hypothesis. The indirect method is to test the hypothesis

$$H_0: \beta_{\cdot 1} = \beta_{\cdot 2} = \cdots = \beta_{\cdot T} \tag{6.39}$$

where

$$\beta_{\cdot t} = \bar{\beta} + u_t \tag{6.40}$$

To perform the test the v_i, u_t, and ε_{it} must be assumed normal. The test statistic is

$$\hat{\beta}_t' Q'(Q \Sigma Q')Q \hat{\beta}_t \tag{6.41}$$

which has a chi-square distribution with T − 1 degrees of freedom under the null hypothesis. The terms in (6.41) are defined as follows:

$$\hat{\beta}_t = \begin{bmatrix} \hat{\beta}_{\cdot 1} \\ \hat{\beta}_{\cdot 2} \\ \cdot \\ \cdot \\ \cdot \\ \hat{\beta}_{\cdot T} \end{bmatrix} \tag{6.42}$$

where $\hat{\beta}_{\cdot t}$ is obtained by applying OLS across individuals at a particular time, t;

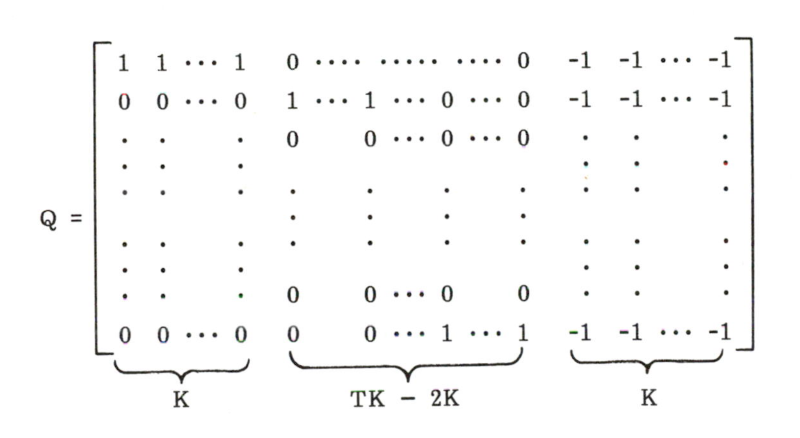

(6.43)

Q is a $(T - 1) \times TK$ matrix;

$$\Sigma = \begin{bmatrix} \Sigma_{11} & \Sigma_{12} & \cdots & \Sigma_{1T} \\ \Sigma_{21} & \Sigma_{22} & \cdots & \Sigma_{2T} \\ \vdots & \vdots & & \vdots \\ \Sigma_{T1} & \Sigma_{T2} & \cdots & \Sigma_{TT} \end{bmatrix} \qquad (6.44)$$

and

$$\Sigma_{tt'} = E(\hat{\beta}_{\cdot t} - \beta_{\cdot t})(\hat{\beta}_{\cdot t} - \beta_{\cdot t'})'$$
$$= (X_t' X_t)^{-1} X_t' [D(X_t \Delta_v X_{t'}') + \sigma^2 I_N] X_{t'} (X_{t'}' X_{t'})^{-1} \qquad (6.45)$$

Hsiao also notes that Δ_v in equation (6.45) can be replaced by a consistent estimator $\hat{\Delta}_v$.

If the null hypothesis in (6.39) is accepted, implying that coefficients do not vary over time, then the hypothesis

$$H_0: \Delta_v = 0 \qquad (6.46)$$

can be indirectly tested by testing

$$H_0: \beta_{1\cdot} = \beta_{2\cdot} = \cdots = \beta_{N\cdot} = \beta \qquad (6.47)$$

where

$$\beta_{i\cdot} = \bar{\beta} + v_i \qquad (6.48)$$

This test was developed by Swamy (1970) and is described in Chapter 5. The reader is referred to equation (5.55) for the test statistic.

Using ML estimation Hsiao proposes a test of the hypothesis

$$H_0: \Delta_v = 0 \qquad (6.49)$$

The test statistic is

$$N \hat{\delta}_v' G^{-1} \hat{\delta}_v \qquad (6.50)$$

where

$$\hat{\delta}_v = (\hat{\delta}_{v11}, \ldots, \hat{\delta}_{vKK})' \tag{6.51}$$

and G^{-1} is the K × K matrix with gkth element

$$\frac{1}{2N} \sum_{i=1}^{N} \text{trace } (G_{ig} \; G_{ik}) \tag{6.52}$$

where

$$G_{ik} = X_{ik} \; X'_{ik} \tag{6.53}$$

The statistic in equation (6.50) will have a chi-square distribution with K + 1 degrees of freedom under the null hypothesis. This test is justified when N is large.

If T is large a similar test can be developed for the hypothesis

$$H_0\colon \; \Delta_u = 0$$

To test the hypothesis

$$H_0\colon \; \bar{\beta} = \bar{\beta}_0 \tag{6.54}$$

Hsiao reccommends the test statistic

$$(\hat{\bar{\beta}} - \bar{\beta}_0)'(X'\hat{\Omega}^{-1}X)(\hat{\bar{\beta}} - \bar{\beta}_0) \tag{6.55}$$

which will have a chi-square distribution with K degrees of freedom when both N and T are large. This statistic will apply when Ω is estimated by either ML or MINQUE as long as N and T are large and some general assumptions hold. The assumption that the v_i, u_t, and ε_{it} be normally distributed is not required in large samples.

6.3 THE SWAMY AND MEHTA MODEL

6.3.1 The Model

Swamy and Mehta (1975a) present the following model:

$$y_{it} = x_{it} \; \beta_{it} \tag{6.56}$$

where

y_{it} is the tth observation of the dependent variable for the ith cross-sectional unit.

x_{it} is a $1 \times K$ vector of observations on the independent variables.

β_{it} is a $K \times 1$ vector of regression coefficients which can be decomposed as

$$\beta_{it} = \bar{\beta} + v_i + u_{it} \tag{6.57}$$

where

$$\bar{\beta} = (\bar{\beta}_1, \bar{\beta}_2, \ldots, \bar{\beta}_K)'$$

is the common mean of the coefficient vectors,

$$v_i = (v_{1i}, v_{2i}, \ldots, v_{Ki})'$$

is constant through time but differs among individuals,

$$u_{it} = (u_{1it}\ u_{2it}, \ldots, u_{Kit})'$$

differs among individuals both at a point in time and through time.

The T observations on each individual can then be written as

$$Y_i = X_i \bar{\beta}_i + D_{x_i} u_i \tag{6.58}$$

where $\bar{\beta}_i = \bar{\beta} + v_i$ is a $K \times 1$ vector,

$$Y_i = \begin{bmatrix} y_{i1} \\ y_{i2} \\ \cdot \\ \cdot \\ \cdot \\ y_{iT} \end{bmatrix} \tag{6.59}$$

$$X_i = \begin{bmatrix} x_{i11} & \cdots & x_{iK1} \\ x_{i12} & \cdots & x_{iK2} \\ \vdots & & \vdots \\ x_{i1T} & \cdots & x_{iKT} \end{bmatrix} \tag{6.60}$$

$$D_{x_i} = \begin{bmatrix} x_{i1} & 0 & \cdots & 0 \\ 0 & x_{i2} & \cdots & 0 \\ \vdots & \vdots & \ddots & \vdots \\ 0 & 0 & \cdots & x_{iT} \end{bmatrix} \tag{6.61}$$

where x_{it} is the tth row of X_i and

$$u_i = (u'_{i1}, u'_{i2}, \ldots, u'_{iT})'$$

is a TK × 1 vector of errors.

Equation (6.58) can be written for all N cross-sectional units as

$$Y = X\bar{\beta} + Dv + D_x u \tag{6.62}$$

$$= X\bar{\beta} + U \tag{6.63}$$

where

$$Y = \begin{bmatrix} Y_1 \\ Y_2 \\ \vdots \\ Y_N \end{bmatrix} \tag{6.64}$$

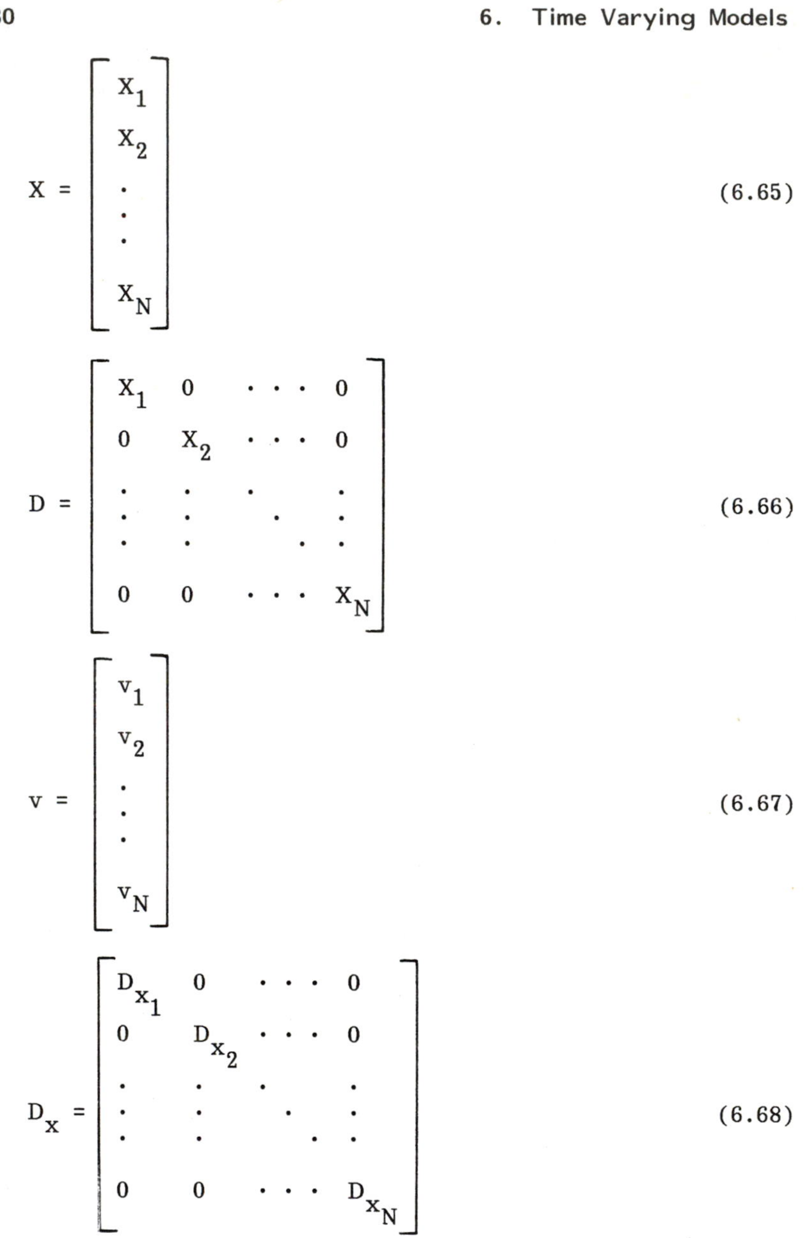

$$X = \begin{bmatrix} X_1 \\ X_2 \\ \vdots \\ X_N \end{bmatrix} \tag{6.65}$$

$$D = \begin{bmatrix} X_1 & 0 & \cdots & 0 \\ 0 & X_2 & \cdots & 0 \\ \vdots & \vdots & \ddots & \vdots \\ 0 & 0 & \cdots & X_N \end{bmatrix} \tag{6.66}$$

$$v = \begin{bmatrix} v_1 \\ v_2 \\ \vdots \\ v_N \end{bmatrix} \tag{6.67}$$

$$D_x = \begin{bmatrix} D_{x_1} & 0 & \cdots & 0 \\ 0 & D_{x_2} & \cdots & 0 \\ \vdots & \vdots & \ddots & \vdots \\ 0 & 0 & \cdots & D_{x_N} \end{bmatrix} \tag{6.68}$$

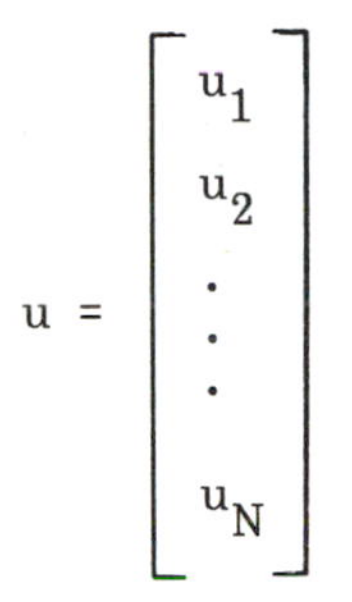

$$u = \begin{bmatrix} u_1 \\ u_2 \\ \cdot \\ \cdot \\ \cdot \\ u_N \end{bmatrix} \tag{6.69}$$

Swamy and Mehta make the following assumptions:

Assumption 6.2A: $E(v_i) = 0$.

Assumption 6.2B: $E(v_i v_i') = \begin{cases} \Delta \text{ if } i = j \\ 0 \text{ otherwise} \end{cases}$

Assumption 6.2C: $E(u_{it}) = 0$.

Assumption 6.2D: $E(u_{it}\, u_{jt'}') = \begin{cases} \Delta_{ii} \text{ if } i = j \quad \text{and} \quad t = t' \\ 0 \text{ otherwise} \end{cases}$

Assumption 6.2E: v_i is independent of u_{it}.

Assumption 6.2F: X_i is fixed in repeated samples with rank $(X_i) = K < T$ and $K < N$.

Assumption 6.2G: $\Delta, \Delta_{11}, \ldots, \Delta_{NN}$ are not necessarily diagonal.

Swamy and Mehta's model differs slightly from that of Hsiao. Hsiao assumes that $u_{it} = u_t$ in equation (6.57), but includes the additive disturbance, ε_{it}, in equation (6.1). Swamy and Mehta omit the additive disturbance since it cannot be distinguished from the u_{it}. Also, Hsiao assumes the Δ_{ii} from assumption 6.2D are equal to Δ_u for all i and that Δ_u and Δ in assumption 6.2B are diagonal. Swamy and Mehta relax these assumptions.

In addition Swamy and Mehta point out that assumption 6.2D can be further relaxed to

$$E(u_{it}\, u_{jt'}') = \begin{cases} \Delta_{ij} & \text{if } t = t' \\ 0 & \text{otherwise} \end{cases}$$

Also note that if all elements of u_{it} except the first one are zero, the Swamy and Mehta model will simplify to the Swamy RCR model discussed in Chapter 5. If $\Delta = 0$ and $\Delta_{ii} = 0$ for all i, the model is the classical pooling model of Chapter 2, that is, the coefficient vectors will be fixed and equal. If $\Delta_{ii} = 0$ for all i and the v_i are treated as fixed parameters the model will be essentially the same as the ANCOVA model of Chapter 4, except that the coefficients will differ only between individuals and not over time.

Singh and Ullah (1974) discuss a model similar to Swamy and Mehta's with less general assumptions.

The variance-covariance matrix of u in equation (6.62) is

$$E(uu') = \Omega = \begin{bmatrix} X_1 \Delta X_1' + \Sigma_{11} & 0 & \cdots & 0 \\ 0 & X_2 \Delta X_2' + \Sigma_{22} & \cdots & 0 \\ \vdots & \vdots & & \vdots \\ 0 & 0 & & X_N \Delta X_N' + \Sigma_{NN} \end{bmatrix} \tag{6.70}$$

where X_i is defined in equation (6.60), and

$$\Sigma_{ii} = D_{x_i} (I_T \otimes \Delta_{ii}) D'_{x_i} \tag{6.71}$$

6.3.2 Estimation Using the Swamy and Mehta Model

The generalized least squares estimator of $\bar{\beta}$ in equation (6.57) is

$$\tilde{\bar{\beta}} = (X'\Omega^{-1}X)^{-1}X'\Omega^{-1}Y \tag{6.72}$$

To construct a feasible estimator of $\bar{\beta}$ the variance-covariance matrix, Ω, must be estimated.

Swamy and Mehta (1977a) derive estimators for Δ and the Δ_{ii} as follows:

Define

$$e_i^* = [I_T - X_i(X_i' \Sigma_{ii}^{*-1} X_i)^{-1} X_i' \Sigma_{ii}^{*-1}] y_i \tag{6.73}$$

and

$$\Sigma_{ii}^* = D_{x_i}(I_T \otimes \Delta_{ii}^*)D_{x_i'} \tag{6.74}$$

where Δ_{ii}^* is a "guessed" value of Δ_{ii}.

Swamy and Mehta (1977a, p. 891) state that a "guessed" value of Δ_{ii} is "any value of Δ_{ii} which is in a close neighborhood of its true value ..." and suggest that the mode of the prior distribution of Δ_{ii} may be the preferred guessed value. If this value is unavailable an iterative procedure may be used where the identity matrix I_k may be used in place of Δ_{ii}^* in the first step and in the second step a consistent estimate of Δ_{ii} is used.

Further define

$$R_i = \Sigma_{ii}^{*-1}[I_T - X_i(X_i' \Sigma_{ii}^{*-1}X_i)^{-1}X_i'\Sigma_{ii}^{*-1}] \tag{6.75}$$

$$u_{it}^* = \Delta_{ii}^* D_{x_{it}'} R_i y_i \tag{6.76}$$

where $D_{x_{it}}$ is the T × K matrix having x_{it}' as its tth row and zeros elsewhere.

Then we have

$$\sum_{t=1}^{T} u_{it}^* u_{it}^{*\prime} = \sum_{t=1}^{T}\sum_{s=1}^{T} \Delta_{ii}^* D_{x_{it}'} R_i D_{x_{is}} \Delta_{ii}^* D_{x_{is}}' R_i' D_{x_{it}} \Delta_{ii}^* \tag{6.77}$$

Define

$$Z_{ii1} = \sum_{t=1}^{T}\sum_{s=1}^{T} (\Delta_{ii}^* D_{x_{it}'} R_i D_{x_{is}} \otimes \Delta_{ii}^* D_{x_{it}'} R_i D_{x_{is}}) \tag{6.78}$$

and δ_{ii1} to be a vector representation of all the K^2 elements δ_{iihl} (h, l = 1,...,K) of Δ_{ii}. The elements of Δ_{ii} are arranged so that the first column of Δ_{ii} appears first, followed by the second column, etc.

Also define

$$\overline{S}_{ii1} = \overline{Z}_{ii1}\delta_{ii2} \tag{6.79}$$

where $\overline{S}_{ii1}$ is the K(K + 1)/2 vector containing the diagonal and twice the below diagonal elements of the matrix in equation (6.77). The order of arrangement of the elements is

$$(S_{111}, 2S_{121},\ldots,2S_{1K1}, S_{122}, 2S_{132},\ldots,2S_{1K2},\ldots,S_{1KK})'$$

The K(K + 1)/2 by K(K + 1)/2 matrix $\bar{Z}_{ii1}$ is obtained by adding the rows of Z_{ii1} corresponding to the hl and lhth $(h \neq l)$ elements of the matrix in equation (6.77), and by adding the columns of Z_{ii1} corresponding to the hl and lhth elements of Δ_{ii}.

The K(K + 1)/2 vector δ_{ii2} contains the diagonal and below diagonal elements of Δ_{ii} arranged as

$$(\delta_{111}, \delta_{121},\ldots,\delta_{1K1}, \delta_{122}, \delta_{132},\ldots,\delta_{1K2},\ldots,\delta_{1KK})' \tag{6.80}$$

The vector S_{ii2} is defined to be a vector representation of all the nonzero elements of the T × T diagonal matrix

$$\sum_{t=1}^{T} D_{x_{it}} u_{it}^* u_{it'}^* D_{x_{it}'} \tag{6.81}$$

and can be written as

$$S_{ii2} = \bar{Z}_{ii2}\, \delta_{ii2} + \eta_{ii2} \tag{6.82}$$

where η_{ii2} is defined as the deviation of S_{ii2} from its expectation $\bar{Z}_{ii2}\, \delta_{ii2}$.

Define Z_{ii2} to be the T × K^2 matrix with tth row

$$\sum_{t'} R_{itt'}{}^2 (x_{it}\, \Delta_{ii}^* x_{it})^2 (x_{it'}' \otimes x_{it'}') \tag{6.83}$$

where $R_{itt'}$ is the tt'th element of R_i.

Define $\bar{Z}_{ii2}$ to be the T × [K(K + 1)/2] matrix obtained by adding the columns of Z_{ii2} corresponding to the hl and lhth $(h \neq l)$ elements of Δ_{ii}.

Swamy and Mehta derive estimators of the elements of Δ and the Δ_{ii} under assumptions 6.2 A-G and the following additional assumptions:

Assumption 6.2H: The rank of $\bar{Z}_{ii2}$ is K(K + 1)/2 for every $T \geq K(K + 1)/2$.

Assumption 6.2I: The vectors v_i and u_{it} are normally distributed.

Assumption 6.2J: The matrix Ω_{ii} is nonsingular for all T > K(K + 1)/2 where

$$E(v_{ii2} v'_{ii2}) = \Omega_{ii} \tag{6.84}$$

Swamy and Mehta (1977a, p. 892) suggest the following estimator of Δ_{ii}:

First estimate δ_{ii2} using either the least squares estimator

$$\hat{\delta}_{ii2} = (\overline{Z}'_{ii2} \overline{Z}_{ii2})^{-1} \overline{Z}'_{ii2} S_{ii2} \tag{6.85}$$

or the GLS estimator

$$\hat{\delta}^*_{ii2} = (\overline{Z}'_{ii2} \Omega_{ii}^{-1} \overline{Z}_{ii2})^{-1} \overline{Z}'_{ii2} \hat{\Omega}_{ii}^{-1} S_{ii2} \tag{6.86}$$

where the tt'th element of $\hat{\Omega}_{ii}$ is equal to

$$2(x'_{it} \Delta^*_{ii} x_{it})^2 (x'_{it'} \Delta^*_{ii} x_{it'})^2 \left(\sum_{s=1}^{T} R_{itt'} R_{it's} x'_{is} \Delta_{ii} x_{is} \right)^2 \tag{6.87}$$

with Δ^*_{ii} equal to the estimate derived by least squares in equation (6.85) or equal to some prior guessed value if the least squares estimate is not positive definite.

The choice of whether to use $\hat{\delta}_{ii2}$ or $\hat{\delta}^*_{ii2}$ depends on efficiency considerations. Swamy and Mehta point out that $\hat{\delta}^*_{ii2}$ should be at least as efficient as $\hat{\delta}_{ii2}$ provided T is large. Assuming this to be the case the following estimator can be constructed:

$$\text{Let } \hat{\Delta}_{ii} = [\hat{\delta}_{iihl}] (h, l = 1, \ldots, K) \tag{6.88}$$

where the $\hat{\delta}_{iihl}$ are values suggested by the estimator $\hat{\delta}^*_{ii2}$.

Swamy and Mehta also construct a truncated estimator of Δ_{ii} to insure that the diagonal elements are positive and the square of the hlth off-diagonal element is less than or equal to the product of the hth and lth diagonal elements. Letting T denote the truncated estimator we have

$$\hat{\Delta}_{iiT} = [d_{iihl}] (h, l = 1, \ldots, K) \tag{6.89}$$

where

$$d_{iihh} = \max(0.01, \hat{\delta}_{iihh})$$

$$d_{iihl}^{2} = \min(\hat{\delta}_{iihl}, d_{iihh}\, d_{iill}) \text{ for } h \neq l = 1, \ldots, K$$

and d_{iihl} with $h \neq l$ is given the same sign as $\hat{\delta}_{iihl}$.

The goal of using the truncated estimator is to reduce the mean squared estimation error and to increase statistical efficiency in small samples. Whether this will occur is a matter for empirical investigation. There is support in simpler situations, however, that truncated estimators can have smaller mean squared errors.

Once the Δ_{ii} have been estimated the matrix Δ can be estimated as

$$\hat{\Delta}_1 = \frac{S_{\hat{\beta}}}{N-1} - \frac{1}{N}\sum_{i=1}^{N}(X_i' \hat{\Sigma}_{ii}^{-1} X_i)^{-1} \tag{6.90}$$

where

$$S_{\hat{\beta}} = \sum_{i=1}^{N} \hat{\beta}_i \hat{\beta}_i' - \frac{1}{N}\sum_{i=1}^{N}\hat{\beta}_i \sum_{i=1}^{N}\hat{\beta}_i' \tag{6.91}$$

$$\hat{\beta}_i = (X_i' \hat{\Sigma}_{ii}^{-1} X_i)^{-1} X_i' \hat{\Sigma}_{ii}^{-1} Y_i \tag{6.92}$$

and $\hat{\Sigma}_{ii}$ is formed by replacing the Δ_{ii} by their estimates.

Alternatively, Δ can be estimated by

$$\hat{\Delta}_2 = \frac{S_{\hat{\beta}}}{N-1} \tag{6.93}$$

Whether $\hat{\Delta}_1$ or $\hat{\Delta}_2$ provides smaller mean squared estimation error is again a matter for empirical investigation.

Once Δ and the Δ_{ii} have been estimated the feasible GLS estimator of $\bar{\beta}$ can be constructed:

$$\hat{\bar{\beta}} = (X'\hat{\Omega}^{-1}X)^{-1}X'\hat{\Omega}^{-1}Y \tag{6.94}$$

where $\hat{\Omega}$ is constructed from Ω in equation (6.70) by replacing Δ and the Δ_{ii} by their estimates.

6.3.3 Inference in Swamy and Mehta's Model

The tests discussed in this section will apply if the sample sizes N and T are large.

To test hypotheses about individual coefficient means, $\bar{\beta}_k$, Swamy and Mehta (1977a) suggest using the ratio of the coefficient estimate to the standard error of the coefficient. These ratios will have a t-distribution with N − 1 degrees of freedom as in Swamy (1970). The standard errors are computed as the square roots of the diagonal elements of

$$(X'\hat{\Omega}^{-1}X)^{-1} \tag{6.95}$$

Certain other tests are also suggested by Swamy and Mehta but are not explicitly stated in their article. Certain test statistics and their distributions will be outlined here.

To test the hypothesis

$$H_0: \bar{\beta} = \bar{\beta}_0$$

where $\bar{\beta}_0$ is a vector of hypothesized values for the coeffficient mean vector $\bar{\beta}$ the following statistic can be used:

$$(\hat{\bar{\beta}} - \bar{\beta}_0)'(X'\hat{\Omega}^{-1}X)(\hat{\bar{\beta}} - \bar{\beta}_0) \tag{6.96}$$

which will have a chi-square distribution with K degrees of freedom when both N and T are large.

Another hypothesis of interest is

$$H_0: \Delta_{ii} = 0 \qquad \text{for all } i \tag{6.97}$$

This is a test of whether the regression coefficients vary over time. If the null hypothesis is accepted, the model reduces to the RCR model of Chapter 5.

Swamy and Mehta (1977a) show that $\sqrt{T}\,(\hat{\delta}^*_{ii2} - \delta_{ii2})$ has a limiting distribution that is normal with mean zero and covariance matrix

$$\left(\lim_{T\to\infty} \frac{1}{T}\, \bar{Z}'_{ii2}\, \Omega_{ii}^{-1} Z_{ii2}\right)^{-1} \tag{6.98}$$

This result can be used to test whether the elements of Δ_{ii} as arranged in δ_{ii2} (see equation (6.80)) are equal to zero. If the hypothesis in (6.97) is true then $\sqrt{T}\,\hat{\delta}^*_{ii2}$ divided by the appropriate elements from the covariance matrix in (6.98) will have a standard normal distribution.

To test hypotheses about Δ, the procedure in Swamy (1971, pp. 318-319) can be used. The estimator $\hat{\Delta}_2$ in equation (6.93) is shown in Swamy and Mehta (1977a) to have a limiting distribution that is Wishart with N − 1 degrees of freedom. To test the hypothesis

$$H_0\colon \Delta = \Delta_0 \tag{6.99}$$

where not all elements of Δ_0 are zero, the test statistic

$$(N-1)\,\frac{\ell'\hat{\Delta}_2\ell}{\ell'\Delta_0\ell} \tag{6.100}$$

can be used. The statistic will have a Chi-square distribution with N − 1 degrees of freedom.

In addition to the tests suggested by Swamy and Mehta, indirect tests for randomness could be developed. These would proceed in a manner similar to those discussed earlier in this chapter for the Hsiao model. The reader is referred to the hypotheses in equations (6.39) and (6.47).

6.4 OTHER MODELS WITH TIME VARYING COEFFICIENTS

Rosenberg (1973b) developed a "stochastically convergent parameter" regression model. In his model the individual coefficients follow a random walk but also have a tendency to converge to a population average.

The model can be written as

$$y_{it} = x_{1it}\beta_{1t} + x_{2it}\beta_{2it} + \varepsilon_{it} \tag{6.101}$$

where

y_{it} is the tth observation of the dependent variable for the ith cross-sectional unit.

x_{1it} is a $1 \times K_1$ vector of observations on K_1 independent variables.

x_{2it} is a $1 \times K_2$ vector of observations on K_2 independent variables.

β_{1t} is a $K_1 \times 1$ vector of parameters which are the same for each cross-sectional unit.

β_{2it} is a $K_2 \times 1$ vector of parameter which vary between cross-sectional units.

ε_{it} is a random disturbance for the ith cross-sectional unit in time period t.

The regression parameters are assumed to evolve as:

$$\beta_{1,t+1} = \beta_{1t} + \gamma_t \tag{6.102}$$

$$\beta_{2,i,t+1} = \bar{\beta}_{2t} + C(\beta_{2it} - \bar{\beta}_{2t}) + \eta_{it} \tag{6.103}$$

for $t = 1,\ldots,T - 1$ and $i = 1,\ldots,N$.

The matrix C is a diagonal matrix with diagonal entries, C_i, which represent convergence rates. That is, C_i is the proportion of the individual difference $\beta_{2it} - \bar{\beta}_{2t}$ which is incorporated into the new coefficient value in period $t + 1$. The value $\bar{\beta}_{2t}$ is the mean value for the cross-sectionally varying coefficient vector in time period t and γ_t and η_{it} are random disturbances.

According to this specification, the cross-sectionally varying coefficients, β_{2it}, tend to converge toward a mean value, $\bar{\beta}_{2t}$, with the rate of convergence depending on the diagonal entries of matrix C. Also, coefficient values may differ randomly over cross-sections or between time periods due to the disturbance η_{it}. The parameters that are fixed cross-sectionally, β_{1t}, are allowed to vary randomly over time.

Rosenberg allows very general assumptions concerning the variance-covariance structure of the random components in the model. Wansbeek and Kapteyn (1978, 1982) examine a model with coefficients that can vary either across individuals or over time. Both fixed effects and random effects versions of the model are considered. With random effects the model can be viewed as a special case of the Hsiao model.

Liu and Hanssens (1981) present a Bayesian analysis of a model with parameters that vary over time only. Harvey (1978) also examines a model with parameter variation over time only.

See Rosenberg (1973a) for an additional survey of time-varying parameter models.

6.5 APPLICATIONS

Few applications of models with time and individual variation in the coefficients have appeared in the literature. This is likely due to the additional complexity of the theory involved in these models and the computational complexity.

Swamy and Mehta (1975a, 1977a) provide examples of the use of their model. Woodland (1983) applied a slightly modified version of the Singh and Ullah (1974) model to estimate the effect of the vintage (time of entry into the labor market) on an individual's lifetime earnings. The Woodland and Singh and Ullah models are less general but similar to the Swamy and Mehta model.

Rosenberg (1974) and Rosenberg and McKibben (1973) discuss the use of the stochastically convergent parameter model of Rosenberg (1973b) to predict systematic risk in common stocks.

7
Models for Discrete Data

7.1 INTRODUCTION

The work discussed in the previous chapters of this book has assumed that the dependent variable could be modeled as a continuous random variable. In certain instances the dependent variable may obviously be discrete. Recently, there has been a considerable amount of research on methods for analyzing models with discrete dependent variables when data on individuals observed over time are available.

Situations where such data may arise include consumer choice among a collection of discrete alternatives over time or analyses of labor force participation. Much of the research in the area of discrete data analysis assumes that panel data are available. The term panel data refers to a collection of observations over time on the same individuals. As in previous chapters, the data are in the form of time series data on a number of individuals. However, here it is typically assumed that the time series will often be very short, perhaps only two observations or at least very few observations over time.

As an example, consider the work of Heckman and Willis (1977) on the dynamics of female labor supply. The dependent variable is defined as either labor force participation ($y = 1$) or nonparticipation ($y = 0$). Their sample consisted of 1583 women observed annually for a 5 year period. Thus, the number of individuals or cross sections was very large, $N = 1583$, but the time series length was short, $T = 5$. When analyzing such data one would desire a methodology whose asymptotic properties depended on large N rather than large T. This is in contrast to the asymptotic properties discussed for many of the methods in previous chapters. Also, the dependent variable is obviously discrete in this example and special methods for dealing with the discrete dependent variable must be utilized.

The work on pooled models for data with discrete dependent variables is relatively new. This chapter is intended to be a brief survey of some of the recent research in the area. The work of a few authors is followed closely in the presentation. The reader is directed to the additional references mentioned for more in-depth work in the area.

For reviews of discrete data models, often called qualitative response models, outside the context of pooled data see Manski and McFadden (1981), McFadden (1974, 1982, 1984) and Amemiya (1981, 1985, pp. 267-354).

7.2 ANALYSIS OF COVARIANCE WITH DISCRETE DEPENDENT VARIABLES

Chamberlain (1980) developed a model based on the analysis of covariance model discussed in Chapter 4 for the case of discrete dependent variables.

The ANCOVA model of Chapter 4 was written as

$$Y = b_0 \iota_{NT} + Wd + \varepsilon \tag{7.1}$$

where b_0 is the constant term, ι_{NT} is an $NT \times 1$ vector of ones, W is the matrix of explanatory variables (excluding the column of ones), d is the vector of slope coefficients to be estimated and ε is a vector of disturbances.

Chamberlain assumes the disturbances can be decomposed as

$$\varepsilon_{it} = c_i + v_{it} \tag{7.2}$$

where the c_i are fixed parameters representing individual effects and the v_{it} are random disturbances.

For an individual the model can be written as (ignoring the overall constant b_0)

$$y_{it} = c_i + w_{it}d + v_{it} \tag{7.3}$$

Now assume that y_{it} is a binary variable such that

$$\text{Prob}(y_{it} = 1 \mid W,d,c) = F(c_i + w_{it}d) \tag{7.4}$$

where $F(\cdot)$ is a cumulative distribution function such as a unit normal or a logistic distribution function.

Chamberlain shows that the usual maximum likelihood (ML) estimator of d is inconsistent in the probability models. He then develops consistent estimators by using a conditional likelihood function. The likelihood function is based on the conditional distribution of the data, conditioning on a set of sufficient statistics for the incidental parameters.

Note that $\Sigma_{t=1}^{T} y_{it}$ is a sufficient statistic for c_i. The conditional log-likelihood function is

$$L = \sum_{i=1}^{N} \ln \frac{\exp(d' \sum_{t=1}^{T} w_{it} y_{it})}{\sum_{\delta \varepsilon B_i} \exp(d' \sum_{t=1}^{T} w_{it} \delta_t)} \tag{7.5}$$

where

$$B_i = \left\{ \delta = (\delta_1, \ldots, \delta_T) \mid \delta_t = 0 \text{ or } 1 \quad \text{and} \quad \sum_{t=1}^{T} \delta_t = \sum_{t=1}^{T} y_{it} \right\} \tag{7.6}$$

L is in the form of a conditional logit log-likelihood function with the alternative set, B_i, varying across the observations. L can be maximized using standard ML logit programs. The matrix

$$J_B^{-1} = \left[\frac{-\partial^2 L}{\partial d \partial d'} \right]^{-1} \tag{7.7}$$

where ∂ denotes the partial derivative, is an appropriate asymptotic covariance matrix for the conditional ML estimator of d provided that J_B/N converges to its expectation.

If y_{it} can take on three values (say, a, b, and c) then

$$\text{Prob}(y_{it} = j) = \frac{\exp(c_{ij} + d'w_{itj})}{\sum_j (\exp(c_{ij} + d'w_{itj}}\qquad \text{for } j = a,b,c \tag{7.8}$$

Define

$$\gamma_{itj} \begin{cases} 1 & \text{if } y_{it} = j \\ 0 & \text{otherwise} \end{cases}$$

Then, conditioning on $\Sigma_{t=1}^{T} \gamma_{itj}$, the conditional log-likelihood function will be

$$L = \sum_i \ln \frac{\exp\left(d'\sum_{t,j} w_{itj}\gamma_{itj}\right)}{\sum_{\delta\varepsilon B_i} \exp\left(d'\sum_{t,j} w_{itj}\delta_{tj}\right)} \tag{7.9}$$

where $B_i = \left\{(\delta_{11}, \ldots, \delta_{TJ}) \mid \delta_{tj} = 0 \text{ or } 1,\ \sum_j \delta_{tj} = 1,\right.$

$$\left.\sum_{t=1}^{T} \delta_{tj} = \sum_{t=1}^{T} \gamma_{itj},\ j=1,\ldots,J\right\} \tag{7.10}$$

Equation (7.9) is in the form of a conditional logit log-likelihood function and can be maximized by standard programs.

The previous analyses have considered the model in equation (7.3) under the assumption that the c_i are fixed effects. Chamberlain also examined the model when the c_i were assumed to be random. This is equivalent to the error components model extended to the case of a discrete dependent variable. Then the likelihood function can be based on the density for y given W, d, and G where G is the distribution function for c:

$$L = \ln \int f(y \mid W,d,c)dG(c|W) \tag{7.11}$$

Chamberlain refers to equation (7.11) as the marginal likelihood function since it is marginal on c rather than conditional on c.

To specify a conditional distribution for c given w that allows for dependence, Chamberlain suggests that the dependence be specified via a linear regression function

$$c_i = \eta' w_i + v_i \tag{7.12}$$

where $w_i' = (w_{i1}', w_{i2}', \ldots, w_{iT}')$ and where v is independent of w.

Letting H represent a family of univariate distribution functions with parameter vector Θ, the log-likelihood function in the random effects case can be written as

$$L = \sum_{i=1}^{N} \ln \int \prod_{t=1}^{T} F(d'w_{it} + \eta' w_i + v)^{y_{it}} [1 - F(d'w_{it} + \eta' w_i + v)]^{1-y_{it}} \, dH(v \mid \Theta) \tag{7.13}$$

Chamberlain notes that this specification leads to a multivariate probit model if F is a standard normal distribution function and H is the distribution function of a $N(0, \sigma_v{}^2)$ random variable.

Chamberlain (1984) extended his work on both the fixed and random effects model and developed a test of whether the random components, c_i, are independent of the explanatory variables.

Hall (1978) discusses a binary probit model with error components (either fixed or random) that can be estimated using the probit routine in the TSCS package.

7.3 DYNAMIC MODELS FOR DISCRETE PANEL DATA

This section discusses models examined primarily by Heckman in several articles. The primary references throughout this section are Heckman (1978 and 1981a) although several related references will be noted throughout. The models examined by Heckman are intended to allow analysis of the structure of discrete choices made over time.

Although work has been done previously on models with discrete dependent variables, the work of Heckman (and others to be mentioned later) adds several new dimensions to possible analyses of discrete data. By considering models which incorporate panel data, the researcher is not limited to examining decision making at a single point in time. The relationship between current choices and past (or future) choices can be considered in the analyses. Several questions, previously difficult or impossible to answer with nondynamic models, can be examined with the model to be discussed in this section. These questions will be mentioned later in the section.

The model to be discussed is based on the notion that discrete outcomes are generated by continuous variables that cross threshholds. For example, in a study of labor force participation, the continuous variable that generates the participation may be the difference between market wages and reservation wages (see Heckman and MaCurdy, 1980).

To develop the model suppose that the sample consists of N individuals observed over T time periods. The presence or absence of some event is recorded for the dependent variable. If the event occurs this is denoted as $y_{it} = 1$ and, if it does not occur, $y_{it} = 0$. Also, note that the event occurs only if some continuous variable, say y^*_{it}, surpasses a certain threshold.

Without loss of generality, the threshold can be assumed to be zero.

If $y^*_{it} \geq 0$, then the event occurs and $y_{it} = 1$.
If $y^*_{it} < 0$, then the event does not occur and $y_{it} = 0$.

Note that y^*_{it} is not observed by the researcher; only the presence or absence of the event is noted. Although y^*_{it} is not observed it can be modeled as the sum of two components:

$$y^*_{it} = w_{it} + \varepsilon_{it} \tag{7.14}$$

where w_{it} represents some function of independent variables and ε_{it} is a random disturbance. w_{it} and ε_{it} may or may not be independent.

Heckman (1981a) assumes that w_{it} consists of purely exogenous variables, Z_{it}, lagged values of the dependent variable and past outcomes of y_{it}. Equation (7.14) can be rewritten as

$$y^*_{it} = Z_{it}\beta + \sum_{j=1}^{\infty} \gamma_{t-j,t}\, y_{i,t-j} + \sum_{j=1}^{\infty} \lambda_{j,t-j} \prod_{\ell=1}^{j} y_{i,t-\ell} + G(L)y^*_{it} + \varepsilon_{it} \tag{7.15}$$

where G(L) is a lag operator of order K so that

$$G(L)y^*_{it} = g_1 L(y^*_{it}) + g_2 L^2(y^*_{it}) + \cdots + g_K L^K(y^*_{it}) \tag{7.16}$$

$$= g_1 y^*_{i,t-1} + g_2 y^*_{i,t-2} + \cdots + g_K y^*_{i,t-K} \tag{7.17}$$

Even though $y^*_{i,t-k}$ is never observed, its effect on current choice can be estimated as pointed out by Heckman (1981a, pp. 145-146).

The initial conditions for $y_{it'}$ and $y^*_{it'}$ fot $t' \leq 0$ are assumed to be determined outside the model. See Heckman (1981b) for a further discussion of the initial condition problem. The disturbance ε_{it} is normally distributed with mean zero.

Stacking the disturbances for individual i into a vector ε_i, Heckman assumes

$$\varepsilon_i \sim N(0, \Sigma) \tag{7.18}$$

where Σ is a $T \times T$ positive definite covariance matrix. It is assumed that $E(\varepsilon_i' \varepsilon_j) = 0$ for $i \neq j$ and that ε_i and Z_{it} are independent.

The term $\Sigma_{j=1}^{\infty} \gamma_{t-j,t} y_{i,t-j}$ represents the effect of past history of the choice process on the current choice. The coefficients, $\gamma_{t-j,t}$, are assumed to depend on the current time period, t, as well as which past time period is being examined, $t - j$.

The term $\Sigma_{j=1}^{\infty} \lambda_{j,t-j} \Pi_{\ell=1}^{j} y_{i,t-\ell}$ represents the cumulative effect on current choices of the most recent experience in a state.

The term $G(L)y^*_{it}$ captures the idea of habit persistence, that is, prior propensities to select a state determine the current probability that a state is occupied.

Now define

$$y_i = (y_{i1}, y_{i2}, \ldots, y_{iT}) \tag{7.19}$$

$$w_{it} = Z_{it}\beta + \sum_{j=1}^{\infty} \gamma_{t-j,t} y_{it} + \sum_{j=1}^{\infty} \lambda_{j,t-j} \prod_{\ell=1}^{\infty} y_{i,t-\ell} + G(L)y^*_{it} \tag{7.20}$$

$$w_i = (w_{i1}, w_{i2}, \ldots, w_{iT}), \text{ and} \tag{7.21}$$

$$C = (\operatorname{diag} \Sigma^{-1})^{1/2} \, \Sigma \, (\operatorname{diag} \Sigma^{-1})^{1/2} \tag{7.22}$$

where diag Σ^{-1} is a vector containing the diagonal elements of Σ^{-1}; C is the correlation matrix derived from the covariance matrix Σ.

The general model can be written as

$$\text{Prob}[y_i \mid Z_{it}, y_{i0}, y_{i,-1}, y_{i,-2,\ldots}, y^*_{i0}, y^*_{i,-1}, y^*_{i,-2}, \ldots]$$
$$= F[W(*)(2y_i - \iota); C(*)(2y_i - \iota)'(2y_i - \iota)] \tag{7.23}$$

where ι is a $1 \times T$ vector of ones, $F(\cdot; C)$ is the cumulative distribution function of a T-variate standardized multivariate normal random variable with correlation matrix C, and (*) denotes the Hadamard product of two vectors or matrices. The Hadamard product of two vectors, a(*)b, is a vector with elements $a_i b_j$. The Hadamard product of two matrices A(*)B is a matrix with elements $a_{ij}b_{ij}$.

Note that the general model represents the probability of occurrence of a particular state given values of the purely exogenous variables and the initial conditions for the process.

The likelihood function is

$$L = \prod_{i=1}^{N} \text{Prob}[y_i \mid Z_{it}, y_{i0}, y_{i,-1}, \ldots, y^*_{i0}, y^*_{i,-1}, \ldots] \tag{7.24}$$

Maximizing ln L produces ML estimators that are consistent, asymptotically normal and efficient.

Heckman discusses several models that have been considered in the literature and demonstrates that they are special cases of his model.

Heckman and Willis (1975), for example, used the following special case of the model in equation (7.15):

$$y^*_{it} = Z_{it}\beta + \varepsilon_{it} \tag{7.25}$$

where

$$\varepsilon_{it} = c_i + u_{it} \tag{7.26}$$

u_{it} is independent and identically distributed with mean zero and variance σ_u^2, and c_i is distributed independently of u_{it}. The error components, c_i, in equation (7.26) are often termed heterogeneity in the applied literature on stochastic processes. Heckman generalized the components of variance scheme to a "one-factor" scheme by writing

$$\varepsilon_{it} = \alpha_t c_i + u_{it} \tag{7.27}$$

where c_i is distributed independently of u_{it}, $E(c_i) = E(u_{it}) = 0$, Var $(u_{it}) = \sigma_{ut}^2$, and $\text{Var}(c_i) = \sigma_c^2$.

The one-factor scheme allows for more general correlation patterns among the unobservables than were previously possible. Thus, the heterogeneity concept is made more general.

More general error components structures can also be hypothesized and the models estimated. For example, Heckman suggests the Balestra and Nerlove (1966) error process

$$\varepsilon_{it} = \rho\varepsilon_{i,t-1} + c_i + u_{it} \tag{7.28}$$

A model with this error structure can be estimated by multivariate probit analysis.

The model in equations (7.25) and (7.26) can also be viewed as a fixed effects model. Heckman noted the advantages of such an approach: 1) computational simplicity, 2) it provides one solution to the initial conditions problem and 3) the researcher estimates the c_i rather than imposing a density. The main disadvantage of the fixed effects approach is that the ML estimator of β is inconsistent.

One of the important uses of the model presented in this section is to determine whether true state dependence is present in a discrete choice situation. Structural state dependence is defined as the structural relationship between discrete outcomes in different periods. For example, it has been noted in various studies (see Heckman and Willis, 1977, for example) that individuals who have experienced a certain event in the past (say, employment) will be more likely to experience that event again in the future than are individuals who have not experienced the event. One explanation for this is that individual preferences are altered as a result of past experience. Past experience has a behavioral effect on the individual and alters future choices. This is true state dependence.

However, it is important to be able to distinguish between true state dependence and "spurious" state dependence. The appearance of state dependence may arise because of improper treatment of unmeasured variables in the model. If individual differences in propensity to experience an event are correlated over time and not controlled for in the model, state dependence may appear to exist. That is, the conditional probability that $y_{it} = 1$ given $y_{i,t-1} = 0$ may not equal the marginal probability that $y_{it} = 1$. If the serially correlated error components are controlled for, such a conditional relationship will not arise. Heckman (1978) has developed tests to distinguish between true and spurious state dependence as did Chamberlain (1984, 1985).

For a detailed discussion of the concepts of heterogeneity and state dependence see Heckman (1981c).

In models with structural state dependence, the presence of a given state in a previous period is important in determining current choices. An alternative to this type of model is one in which habit persistence is present. In models with habit persistence, it is not the presence or absence of a given state that is important in determining current choice; rather, it is the individual's previous propensities to select a given state that help determine the current probability that a state is occupied. The simplest model representing habit persistence might be written

$$y_{it}^* = G(L)y_{it}^* + \varepsilon_{it} \tag{7.29}$$

where G(L) was defined as the general lag operator. Thus, values of y_{it}^* in previous time periods are important in determining the current state. Again, as pointed out by Heckman, the coefficients of G(L) can be estimated by multivariate probit analysis provided $T \geq K$ and the initial conditions are specified.

The ε_{it} must be independent and identically distributed. Otherwise the model is not identified. This is the discrete analog of the problem in time series regression analysis of estimating a distributed lag model with serially correlated disturbances.

As mentioned previously, the y_{it}^* are never actually observed. However, their effect on choice of the current state can still be estimated. Also, hypothesis tests can be used to distinguish between models with habit persistence and models with structural state dependence.

Applications include Heckman and Willis (1975, 1977). See also Shapiro and Mott (1978).

HotzTran is a statistical package that includes options to estimate discrete choice models with panel data. HotzTran is available from CERA Economic Consultants, Inc.

7.4 FURTHER RESEARCH

A new area of recent research is that of duration analysis. Heckman and Singer (1984a) provide a review of much of the research on econometric duration analysis. They suggest that, in analyzing discrete choices made over time there are two arguments which favor such duration analysis models: 1) the models operate in continuous time, which is a more realistic setting than the modeling of decisions in a discrete time setting, since there is no natural time unit (days, weeks, etc.) in which economic agents make decisions and 2) even if there were such a natural time unit it might not correspond to the units in which data are gathered.

The duration models operate in continuous time although they are still used to analyze discrete choice processes. These models will not be discussed further in this book. They are the subject of much current research, however. The interested reader is referred to the Heckman and Singer (1984a) article as well as the following related articles: Flinn and Heckman (1982a, 1982b, 1983a, 1983b) and Heckman and Singer (1982, 1984b, 1984c, 1985).

Daganzo and Sheffi (1982) extend the work of Heckman (1981a) to include multinomial choice models. In their model, each individual is faced with J different alternatives (rather than two as discussed previously) and is still observed over T time periods. They show that the problem can be estimated using multinomial probit. The complexity of the estimation process increases with the product of the number of alternatives and the number of time periods. Their model takes account of both serial correlation and state dependence as does Heckman's (1981a) binary model. Johnson and Hensher (1982) applied the multinomial probit model to data on commuters' choice of the mode of transportation used.

Avery, Hansen, and Hotz (1983) consider a simpler version of Heckman's (1978, 1981a) model that excludes state dependence and lagged dependent variables. They note that ML estimation of the model may result in inconsistent parameter estimates if the variance-covariance matrix is incorrectly specified. Estimators are developed that are computationally practical and are consistent and asymptotically normal under certain misspecifications of the covariance matrix. When no misspecification occurs they are less efficient than the ML estimators.

Moffitt (1984) applies procedures related to those proposed by Heckman to a simultaneous equation model for panel data.

See also Davies and Crouchley (1984) and Heckman and Borjas (1980).

8
Guidelines for Model Choice

8.1 INTRODUCTION

In this chapter an attempt is made to provide some suggestions in choosing an appropriate model for pooled cross-sectional and time series data. The chapter is divided into two sections. In Section 8.2 the results of a simulation study are reported. The simulation provides results concerning primarily, the RCR techniques of Chapter 5, although some comparison with the CP techniques of Chapter 2 is made. The results are for a small sample situation (T = 20 and N = 20). Little is known of the behavior of RCR estimators or tests in small samples and exact finite sample results are difficult, if not impossible, to derive. The simulation thus provides some insight into how well the RCR model performs in small samples.

In Section 8.3 some guidelines for choosing an appropriate model are provided. Though not intended to cover every case, it is hoped that the suggestions might at least aid the researcher in adopting an adequate model for his or her situation.

8.2 A SIMULATION STUDY OF RCR PROCEDURES

A Monte Carlo simulation was conducted to study the behavior of certain estimators and tests in small samples. The simulation was designed primarily to investigate estimation and hypothesis tests for the random coefficient regression (RCR) model of Chapter 5.

Comparisons will be made, however, to the behavior of the classical pooling (CP) estimator of Chapter 2.

The model studied was as follows:

$$y_{it} = \beta_{0i} + \beta_{1i} x_{it} + \varepsilon_{it} \tag{8.1}$$

for i = 1,...,N; t = 1,...,T.

The independent variable and the disturbances in the equation were generated according to a number of specifications in order to provide a variety of conditions under which estimation could take place. The parameters β_{0i} and β_{1i} were generated as both random and fixed. Thus, the procedures discussed for the RCR model in Chapter 5 can be examined both when the assumptions of the model are met and when they are not. The settings of the model are described below and results of the simulation are discussed. Suggestions for more detailed examination of the RCR estimation and testing procedures are also discussed.

To perform the simulation, the model in equation (8.1) was generated as follows:

1. The values of the independent variable x_{it}, were generated as independent normally distributed random variates with mean μ_X and standard deviation σ_X. The values of x_{it} were allowed to differ for each cross-sectional unit; however, once generated for all N cross-sectional units the values were held fixed over all Monte Carlo trials. The value of μ_X was set equal to zero. Values of σ_X used were 1.0 and 10.0 to examine the effect of increasing variability in the independent variable.

2. The disturbances, ε_{it}, were generated as independent normally distributed random variates, independent of the x_{it} values, with mean zero and standard deviation σ_ε. The disturbances were allowed to differ for each cross-sectional unit on a given Monte Carlo trial and were allowed to differ between trials. The standard deviation of the disturbances was set equal to either 1, 10, or 20 and held fixed for each cross-sectional unit. Also, to allow σ_ε to differ for each cross-section, experiments were performed where σ_ε was chosen randomly for each cross-section from a uniform distribution with mean 10 or with mean 20. The uniform distributions used allowed σ_ε to vary over the range (0,20) or (0,40), respectively.

3. The values of N and T were each fixed at 20. This value was chosen as representative of a small sample situation in both the individual and time dimensions. Little is known of the properties of the RCR estimators or the behavior of the test statistics in such small sample cases. All the known analytic results are derived assuming T is large ($T \to \infty$) or N is large ($N \to \infty$).

4. The parameters, β_{0i} and β_{1i}, were set at several different values to allow study of the estimators under conditions where the model was both properly and improperly specified. Also, tests of hypotheses were examined to determine the observed level of significance and to obtain an idea of the power of the tests. The five different combinations of β_{0i} and β_{1i} used are detailed in Table 8.1 by giving the means and standard deviations of the coefficients. Note that a standard deviation of zero simply means that the coefficient is fixed and equal over all cross-sectional units. The combinations of β_{0i} and β_{1i} provided five different model specifications labeled A through E.

Models A, B, and C are random coefficient models. Thus the RCR model would be properly specified for these coefficient combinations. The coefficient means differ to allow an assessment of the observed level of significance and the power of the tests. Models D and E have either one or two fixed coefficients and use of the RCR model would be a misspecification. The mixed RCR model would be appropriate for D and the CP model for E. However, these models will be estimated using RCR in order to study the behavior of the coefficient mean estimator under misspecification of the model and to study the behavior of the tests for randomness of coefficients.

There were a total of fifty experimental settings for the simulation: two independent variable generating schemes, five schemes for generating disturbances and five settings for the regression coefficients. For each of the experimental settings 1,000 Monte Carlo trials were used and results were recorded in Tables 8.2 through 8.11, with each table consisting of five panels numbered I through V. Each of the tables provides the results for a particular scheme of generation of the regression coefficients and the independent variable, while each panel represents one of the five settings of the disturbance standard deviation.

TABLE 8.1 Values of Coefficient Means and Standard Deviations Used in the Simulation

Model	$\bar{\beta}_0$	$\sigma_{\beta 0}$	$\bar{\beta}_1$	$\sigma_{\beta 1}$
A	1	1	1	1
B	0	1	1	1
C	0	1	0	1
D	1	0	1	1
E	1	0	1	0

TABLE 8.2 Results of RCR Estimation when $\beta_{0i} \sim N(0,1)$, $\beta_{1i} \sim N(0,1)$, $X_{it} \sim N(0,1)$ and $\varepsilon_{it} \sim N(0,\sigma_\varepsilon)$

		Estimated Mean	Percent Rejections $H_0: \beta_k = 0$	Percent Coefficients Contained 95% Confidence Interval	Estimated Standard Deviation	Percent Rejections $H_0: \sigma^2_{\beta_k} = 0$	Percent Negative Variance Estimates	Percent Rejections Minus Negative Estimate Rejections
I. $\sigma_\epsilon = 1.0$	$\bar{\beta}_0$	-0.002	5.5	94.5	0.985	100.0	0.0	100.0
	$\bar{\beta}_1$	-0.002	5.5	94.5	0.988	100.0	0.0	100.0
II. $\sigma_\epsilon = 10.0$	$\bar{\beta}_0$	-0.007	6.2	93.8	0.969	100.0	0.0	100.0
	$\bar{\beta}_1$	-0.003	7.0	92.8	0.937	100.0	0.2	99.8
III. $\sigma_\epsilon = 20.0$	$\bar{\beta}_0$	-0.045	14.9	82.7	0.916	99.0	3.5	96.0
	$\bar{\beta}_1$	0.068	17.5	72.8	0.733	97.9	16.6	82.2
IV. $\sigma_\epsilon \sim U(0,20)$	$\bar{\beta}_0$	0.001	6.3	93.6	0.962	100.0	0.3	99.7
	$\bar{\beta}_1$	0.009	7.4	91.2	0.904	100.0	1.9	98.1
V. $\sigma_\epsilon \sim U(0, 40)$	$\bar{\beta}_0$	0.045	17.3	78.4	0.901	99.6	5.9	93.8
	$\bar{\beta}_1$	-0.136	22.4	58.1	0.667	99.2	27.9	71.8

TABLE 8.3 Results of RCR Estimation When $\beta_{0i} \sim N(0,1)$, $\beta_{1i} \sim N(0,1)$, $X_{it} \sim N(0,10)$ and $\varepsilon_{it} \sim N(0,\sigma_\varepsilon)$

		Estimated Mean	Percent Rejections $H_0: \beta_k = 0$	Percent Coefficients Contained 95% Confidence Interval	Estimated Standard Deviation	Percent Rejections $H_0: \sigma^2_{\beta_k} = 0$	Percent Negative Variance Estimates	Percent Rejections Minus Negative Estimate Rejections
I. $\sigma_\epsilon = 1.0$	$\bar{\beta}_0$	-0.002	5.5	94.5	0.985	100.0	0.0	100.0
	$\bar{\beta}_1$	-0.004	4.7	95.3	0.989	100.0	0.0	100.0
II. $\sigma_\epsilon = 10.0$	$\bar{\beta}_0$	-0.007	6.2	93.9	0.969	100.0	0.0	100.0
	$\bar{\beta}_1$	-0.003	7.0	95.1	0.989	100.0	0.0	100.0
III. $\sigma_\epsilon = 20.0$	$\bar{\beta}_0$	-0.043	9.3	89.1	0.916	99.0	3.5	96.0
	$\bar{\beta}_1$	-0.004	5.2	94.8	0.988	100.0	0.0	100.0
IV. $\sigma_\epsilon \sim U(0,20)$	$\bar{\beta}_0$	0.001	5.6	94.2	0.962	100.0	0.3	99.7
	$\bar{\beta}_1$	-0.004	5.1	94.9	0.988	100.0	0.0	100.0
V. $\sigma_\epsilon \sim U(0, 40)$	$\bar{\beta}_0$	0.014	10.5	85.9	0.901	99.6	5.9	93.8
	$\bar{\beta}_1$	-0.004	5.0	95.0	0.986	100.0	0.0	100.0

TABLE 8.4 Results of RCR Estimation When $\beta_{0i} \sim N(1,1)$, $\beta_{1i} \sim N(1,1)$, $X_{it} \sim N(0,1)$ and $\varepsilon_{it} \sim N(0,\sigma_\varepsilon)$

		Estimated Mean	Percent Rejections $H_0: \beta_k = 0$	Percent Coefficients Contained 95% Confidence Interval	Estimated Standard Deviation	Percent Rejections $H_0: \sigma^2_{\beta_k} = 0$	Percent Negative Variance Estimates	Percent Rejections Minus Negative Estimate Rejections
I. $\sigma_\epsilon = 1.0$	$\bar{\beta}_0$	0.998	98.5	94.5	0.985	100.0	0.0	100.0
	$\bar{\beta}_1$	0.998	98.1	94.5	0.988	100.0	0.0	100.0
II. $\sigma_\epsilon = 10.0$	$\bar{\beta}_0$	0.993	96.1	93.8	0.969	100.0	0.0	100.0
	$\bar{\beta}_1$	0.997	95.8	92.8	0.937	100.0	0.2	99.8
III. $\sigma_\epsilon = 20.0$	$\bar{\beta}_0$	0.955	85.9	82.7	0.916	99.0	3.5	96.0
	$\bar{\beta}_1$	1.069	79.8	72.8	0.733	97.9	16.6	82.2
IV. $\sigma_\epsilon \sim U(0,20)$	$\bar{\beta}_0$	1.001	96.6	93.6	0.962	100.0	0.3	100.0
	$\bar{\beta}_1$	1.009	94.8	91.6	0.904	100.0	1.9	100.0
V. $\sigma_\epsilon \sim U(0, 40)$	$\bar{\beta}_0$	1.045	85.3	78.4	0.901	99.6	5.9	93.8
	$\bar{\beta}_1$	0.864	70.7	58.1	0.667	99.2	27.9	71.8

TABLE 8.5 Results of RCR Estimation When $\beta_{0i} \sim N(1,1)$, $\beta_{1i} \sim N(1,1)$, $X_{it} \sim N(0,10)$ and $\varepsilon_{it} \sim N(0,\sigma_\varepsilon)$

		Estimated Mean	Percent Rejections $H_0:\beta_k = 0$	Percent Coefficients Contained 95% Confidence Interval	Estimated Standard Deviation	Percent Rejections $H_0:\sigma^2_{\beta_k} = 0$	Percent Negative Variance Estimates	Percent Rejections Minus Negative Estimate Rejections
I. $\sigma_\epsilon = 1.0$	$\bar{\beta}_0$	0.998	98.5	94.5	0.985	100.0	0.0	100.0
	$\bar{\beta}_1$	0.996	98.9	95.4	0.989	100.0	0.0	100.0
II. $\sigma_\epsilon = 10.0$	$\bar{\beta}_0$	0.993	96.3	93.9	0.969	100.0	0.0	100.0
	$\bar{\beta}_1$	0.996	98.9	95.1	0.989	100.0	0.0	100.0
III. $\sigma_\epsilon = 20.0$	$\bar{\beta}_0$	0.957	89.1	89.1	0.916	99.0	3.5	96.0
	$\bar{\beta}_1$	0.996	98.7	94.8	0.988	100.0	0.0	100.0
IV. $\sigma_\epsilon \sim U(0,20)$	$\bar{\beta}_0$	1.001	96.5	94.2	0.962	100.0	0.3	99.7
	$\bar{\beta}_1$	0.996	99.1	94.9	0.988	100.0	0.0	100.0
V. $\sigma_\epsilon \sim U(0, 40)$	$\bar{\beta}_0$	1.014	88.5	85.9	0.901	99.6	5.9	93.8
	$\bar{\beta}_1$	0.996	99.0	95.0	0.986	100.0	0.0	100.0

TABLE 8.6 Results of RCR Estimation When $\beta_{0i} \sim N(0,1)$, $\beta_{1i} \sim N(1,1)$, $X_{it} \sim N(0,1)$ and $\varepsilon_{it} \sim N(0,\sigma_\varepsilon)$

		Estimated Mean	Percent Rejections $H_0: \beta_k = 0$	Percent Coefficients Contained 95% Confidence Interval	Estimated Standard Deviation	Percent Rejections $H_0: \sigma^2_{\beta_k} = 0$	Percent Negative Variance Estimates	Percent Rejections Minus Negative Estimate Rejections
I. $\sigma_\epsilon = 1.0$	$\bar{\beta}_0$	-0.002	5.5	94.5	0.985	100.0	0.0	100.0
	$\bar{\beta}_1$	0.998	98.1	94.5	0.988	100.0	0.0	100.0
II. $\sigma_\epsilon = 10.0$	$\bar{\beta}_0$	-0.007	6.2	93.8	0.969	100.0	0.0	100.0
	$\bar{\beta}_1$	0.997	95.8	92.8	0.937	100.0	0.2	99.8
III. $\sigma_\epsilon = 20.0$	$\bar{\beta}_0$	-0.045	14.9	82.7	0.916	99.0	3.5	96.0
	$\bar{\beta}_1$	1.068	79.8	72.8	0.733	97.9	16.6	82.2
IV. $\sigma_\epsilon \sim U(0,20)$	$\bar{\beta}_0$	0.001	6.3	93.6	0.962	100.0	0.3	99.7
	$\bar{\beta}_1$	1.009	94.8	91.2	0.904	100.0	1.9	98.1
V. $\sigma_\epsilon \sim U(0, 40)$	$\bar{\beta}_0$	0.045	17.3	78.4	0.901	99.6	5.9	93.8
	$\bar{\beta}_1$	0.364	70.7	58.1	0.667	99.2	27.9	71.8

TABLE 8.7 Results of RCR Estimation When $\beta_{0i} \sim N(0,1)$, $\beta_{1i} \sim N(1,1)$, $X_{it} \sim N(0,10)$ and $\varepsilon_{it} \sim N(0,\sigma_\varepsilon)$

		Estimated Mean	Percent Rejections $H_0: \beta_k = 0$	Percent Coefficients Contained 95% Confidence Interval	Estimated Standard Deviation	Percent Rejections $H_0: \sigma^2_{\beta_k} = 0$	Percent Negative Variance Estimates	Percent Rejections Minus Negative Estimate Rejections
I. $\sigma_\epsilon = 1.0$	$\bar{\beta}_0$	-0.002	5.5	94.5	0.985	100.0	0.0	100.0
	$\bar{\beta}_1$	0.996	98.9	95.3	0.989	100.0	0.0	100.0
II. $\sigma_\epsilon = 10.0$	$\bar{\beta}_0$	-0.007	6.1	93.9	0.969	100.0	0.0	100.0
	$\bar{\beta}_1$	0.996	98.9	95.1	0.989	100.0	0.0	100.0
III. $\sigma_\epsilon = 20.0$	$\bar{\beta}_0$	-0.043	9.3	89.1	0.916	99.0	3.5	96.0
	$\bar{\beta}_1$	0.996	98.7	94.8	0.988	100.0	0.0	100.0
IV. $\sigma_\epsilon \sim U(0,20)$	$\bar{\beta}_0$	0.001	5.6	94.2	0.962	100.0	0.3	99.7
	$\bar{\beta}_1$	0.996	99.1	94.9	0.988	100.0	0.0	100.0
V. $\sigma_\epsilon \sim U(0, 40)$	$\bar{\beta}_0$	0.014	10.5	85.9	0.901	99.6	5.9	93.8
	$\bar{\beta}_1$	0.996	99.0	95.0	0.986	100.0	0.0	100.0

TABLE 8.8 Results of RCR Estimation When $\beta_{0i} = 1.0$, $\beta_{1i} \sim N(1,1)$, $X_{it} \sim N(0,1)$ and $\varepsilon_{it} \sim N(0,\sigma_\varepsilon)$

		Estimated Mean	Percent Rejections $H_0: \bar{\beta}_k = 0$	Percent Coefficients Contained 95% Confidence Interval	Estimated Standard Deviation	Percent Rejections $H_0: \sigma^2_{\beta_k} = 0$	Percent Negative Variance Estimates	Percent Rejections Minus Negative Estimate Rejections
I. $\sigma_\epsilon = 1.0$	$\bar{\beta}_0$	0.993	97.9	85.7	0.049	12.9	55.0	12.9
	$\bar{\beta}_1$	0.997	98.3	94.3	0.988	100.0	0.0	100.0
II. $\sigma_\epsilon = 10.0$	$\bar{\beta}_0$	1.357	80.7	56.5	0.078	4.7	59.4	3.7
	$\bar{\beta}_1$	0.853	94.4	89.0	0.937	100.0	0.2	99.8
III. $\sigma_\epsilon = 20.0$	$\bar{\beta}_0$	5.876	70.0	38.1	0.107	2.1	62.0	1.1
	$\bar{\beta}_1$	2.048	77.4	67.4	0.733	97.9	16.6	82.2
IV. $\sigma_\epsilon \sim U(0,20)$	$\bar{\beta}_0$	1.043	76.0	51.7	0.094	6.1	55.6	4.7
	$\bar{\beta}_1$	1.018	92.4	86.1	0.904	100.0	1.9	98.1
V. $\sigma_\epsilon \sim U(0, 40)$	$\bar{\beta}_0$	0.560	61.8	61.8	0.129	3.2	59.7	1.5
	$\bar{\beta}_1$	0.680	69.0	69.0	0.667	99.2	27.9	71.8

TABLE 8.9 Results of RCR Estimation When $\beta_{0i} = 1.0$, $\beta_{1i} \sim N(1,1)$, $X_{it} \sim N(0,10)$ and $\varepsilon_{it} \sim N(0,\sigma_\varepsilon)$

		Estimated Mean	Percent Rejections $H_0: \bar{\beta}_k = 0$	Percent Coefficients Contained 95% Confidence Interval	Estimated Standard Deviation	Percent Rejections $H_0: \sigma^2_{\beta_k} = 0$	Percent Negative Variance Estimates	Percent Rejections Minus Negative Estimate Rejections
I. $\sigma_\epsilon = 1.0$	$\bar{\beta}_0$	0.970	97.6	85.5	0.049	12.9	55.0	12.9
	$\bar{\beta}_1$	0.995	98.7	95.2	0.989	100.0	0.0	100.0
II. $\sigma_\epsilon = 10.0$	$\bar{\beta}_0$	0.993	81.1	58.2	0.078	4.7	59.4	3.7
	$\bar{\beta}_1$	0.993	98.8	95.0	0.989	100.0	0.0	100.0
III. $\sigma_\epsilon = 20.0$	$\bar{\beta}_0$	0.975	67.3	44.4	0.107	2.1	62.0	1.1
	$\bar{\beta}_1$	1.023	98.4	94.6	0.988	100.0	0.0	100.0
IV. $\sigma_\epsilon \sim U(0,20)$	$\bar{\beta}_0$	1.060	79.0	57.5	0.094	6.1	55.6	4.7
	$\bar{\beta}_1$	0.993	98.8	94.6	0.988	100.0	0.0	100.0
V. $\sigma_\epsilon \sim U(0, 40)$	$\bar{\beta}_0$	1.121	61.9	42.5	0.078	4.7	59.4	3.7
	$\bar{\beta}_1$	0.997	99.0	94.5	0.989	100.0	0.0	100.0

TABLE 8.10 Results of RCR Estimation When $\beta_{0i} = 1.0$, $\beta_{1i} = 1.0$, $X_{it} \sim N(0,1)$ and $\varepsilon_{it} \sim N(0,\sigma_\varepsilon)$

		Estimated Mean	Percent Rejections $H_0: \bar{\beta}_k = 0$	Percent Coefficients Contained 95% Confidence Interval	Estimated Standard Deviation	Percent Rejections $H_0: \sigma^2_{\beta_k} = 0$	Percent Negative Variance Estimates	Percent Rejections Minus Negative Estimate Rejections
I. $\sigma_\epsilon = 1.0$	$\bar{\beta}_0$	0.985	97.7	78.9	0.049	12.9	55.0	12.9
	$\bar{\beta}_1$	0.988	93.6	73.8	0.059	15.2	54.4	14.7
II. $\sigma_\epsilon = 10.0$	$\bar{\beta}_0$	0.894	82.2	50.6	0.078	4.7	59.4	3.7
	$\bar{\beta}_1$	0.948	58.8	31.0	0.072	2.4	75.4	1.1
III. $\sigma_\epsilon = 20.0$	$\bar{\beta}_0$	1.286	72.8	44.0	0.107	2.1	62.0	1.1
	$\bar{\beta}_1$	4.508	47.2	25.3	0.121	0.8	77.8	0.1
IV. $\sigma_\epsilon \sim U(0,20)$	$\bar{\beta}_0$	0.435	78.6	50.4	0.094	6.1	55.6	4.7
	$\bar{\beta}_1$	1.616	51.1	26.5	0.099	5.6	72.7	3.6
V. $\sigma_\epsilon \sim U(0, 40)$	$\bar{\beta}_0$	1.143	66.6	41.5	0.129	3.2	59.7	1.5
	$\bar{\beta}_1$	0.977	38.7	23.5	0.179	2.0	74.8	0.5

TABLE 8.11 Results of RCR Estimation When $\beta_{0i} = 1.0$, $\beta_{1i} = 1.0$, $X_{it} \sim N(0,10)$ and $\varepsilon_{it} \sim N(0,\sigma_\varepsilon)$

		Estimated Mean	Percent Rejections $H_0: \bar{\beta}_k = 0$	Percent Coefficients Contained 95% Confidence Interval	Estimated Standard Deviation	Percent Rejections $H_0: \sigma^2_{\beta_k} = 0$	Percent Negative Variance Estimates	Percent Rejections Minus Negative Estimate Rejections
I. $\sigma_\epsilon = 1.0$	$\bar{\beta}_0$	0.983	97.6	78.9	0.049	12.9	55.0	12.9
	$\bar{\beta}_1$	0.999	93.9	73.7	0.006	15.2	54.4	14.7
II. $\sigma_\epsilon = 10.0$	$\bar{\beta}_0$	0.869	82.3	50.6	0.078	4.7	59.4	3.7
	$\bar{\beta}_1$	0.993	60.5	31.0	0.007	2.4	75.4	1.1
III. $\sigma_\epsilon = 20.0$	$\bar{\beta}_0$	1.618	72.8	44.0	0.107	2.1	62.0	1.1
	$\bar{\beta}_1$	1.843	51.1	25.3	0.012	0.8	77.8	0.1
IV. $\sigma_\epsilon \sim U(0,20)$	$\bar{\beta}_0$	-3.962	78.6	50.4	0.094	6.1	55.6	4.7
	$\bar{\beta}_1$	1.425	52.5	26.5	0.010	5.6	72.7	3.6
V. $\sigma_\epsilon \sim U(0, 40)$	$\bar{\beta}_0$	1.144	66.6	41.5	0.129	3.2	59.7	1.5
	$\bar{\beta}_1$	0.994	43.6	23.5	0.018	2.0	74.8	0.5

In the tables of results, several estimators and test statistics are of interest. Tables 8.2 through 8.11 are set up to show the following information:

1. The coefficient mean estimators (or the estimators of the fixed coefficients), $\hat{\bar{\beta}}_0$ and $\hat{\bar{\beta}}_1$, that are computed as in equation (5.14). The values shown in the first column of each panel of each table are the averages over all 1,000 Monte Carlo trials at a particular setting.

2. The second column shows the percentage of rejections of the null hypothesis H_0: $\bar{\beta}_k = 0$ for k = 0 and 1. The test uses the t-statistic computed in equation (5.43). A nominal 5% level of significance was used so the expected percentage of rejections whenever the null hypothesis is true is 5%.

3. The percentage of time a 95% cofidence interval estimate of $\bar{\beta}_k$ contained the true value of the coefficient is reported in column three. The confidence interval is constructed as

$$\hat{\bar{\beta}}_k \pm t \text{ se } (\hat{\bar{\beta}}_k) \tag{8.2}$$

where se $(\hat{\bar{\beta}}_k)$ is the square root of the kth diagonal element of the variance-covariance matrix given in equation (5.41).

4. The estimated standard deviation of each coefficient, $\hat{\sigma}_{\beta_k}$, averaged over 1,000 trials, is shown in the fourth column. The estimates are computed as the square roots of the diagonal elements in equation (5.11).

5. The fifth column records the percentage of rejections of the null hypothesis H_0: $\sigma^2_{\beta_k} = 0$, for k = 0 and 1, at a nominal 5% level of significance. The chi-squared statistic in equation (5.55) is used to perform the test.

6. As noted in Chapter 5, it is possible to obtain negative estimates of the coefficient variances, $\sigma^2_{\beta_k}$, when equation (5.11) is used to compute the variance-covariance matrix. The sixth column records the percentages of negative variance estimates obtained.

7. The seventh column is the percentage of rejections of the null hypothesis H_0: $\sigma^2_{\beta_k} = 0$ minus the number of times these rejections resulted when a negative variance estimate was observed. The purpose of this column will be explained more fully later in this section.

As a guide to interpreting the tables consider Table 8.2, Panel I. The averages of the coefficient estimates over all 1,000 Monte Carlo trials were both equal to −0.002. Note that the true coefficient values were zero. Also note that 94.5% of the confidence intervals constructed for each of the coefficients contained the true coefficient value (zero). These percentages are very close to the expected 95% enclosure rate. In testing the hypothesis H_0: $\bar{\beta}_k = 0$ the null hypothesis was rejected in 5.5% of all 1,000 trials for each coefficient mean. Since the null hypothesis was true under this setting a rejection rate close to the nominal level of significance, 5%, would be expected. The fourth column shows the average of all 1,000 estimates of the coefficient standard deviation, $\sigma_{\hat{\beta}_k}$. The true standard deviations were equal to 1.0. Column five shows the percentage of rejections of the null hypothesis H_0: $\sigma^2_{\beta_k} = 0$. Since the null hypothesis is false in this setting, the percentage can be used to assess the power of the test. The null hypothesis was rejected in all cases under this setting. To obtain a more complete picture of the power of the test, a more extensive simulation using different values of σ_{β_k} would be necessary. However, the result in panel I does suggest that the test for randomness is performing as designed even in this small sample case.

Column six provides the percentage of negative variance estimates for each coefficient. In this case there were none. In order to consider this percentage more fully consider panel III in the same table. Here the standard deviation of the disturbances is 20.0 rather than 1.0; thus the dependent variable values have a much greater variation, and less of the total variation in Y will be explained by the regressions. As can be seen in this case the test of H_0: $\bar{\beta}_k = 0$ does not perform as well as in panel I. The added variability has increased the observed type I error percentages to 14.9% and 17.5%. The percentage of rejections of the null hypotheses H_0: $\sigma^2_{\beta_k} = 0$ is still very high. Note, however, that in 3.5% of the trials a negative estimate of $\sigma^2_{\beta_0}$ occurred and in 16.6% of the trials a negative estimate of $\sigma^2_{\beta_1}$ occurred.

In column seven, the percentage of cases where negative variance estimates occurred and the test result was to reject, is subtracted from the percent rejections of H_0: $\sigma^2_{\beta_k} = 0$ in column five. The motivation for this is as follows: In a model where disturbances are properly specified, the appearance of a negative variance estimate is likely the result of a fixed coefficient being treated as though it were random. The results of hypothesis tests in cases where negative variance estimates occur may not be reliable (See Dielman (1980, pp. 40-51)). Thus, in column seven, negative variance estimates are treated the same as a decision to accept the null hypothesis that the coefficient is fixed.

Examining Tables 8.2 through 8.11 several observations concerning the RCR estimators and the test statistics can be made:

1. The RCR coefficient mean estimator performs well when the coefficients are random, even though the sample sizes are small (N = 20 and T = 20).

2. When one or both of the coefficients is fixed, the RCR estimator does not perform as well as might be expected.

3. When one or both of the coefficients is fixed, a large number of negative variance estimates occurs as suggested in Dielman (1980, pp. 40-51). Thus, the appearance of negative variance estimates would suggest the possibility that the coefficient be treated as fixed.

4. The test for randomness performs well overall. The test produces a high percentage of rejections of the hypothesis H_0: $\sigma^2_{\beta_k} = 0$ when the coefficients are random and a low percentage when the null hypothesis is true. The adjustment made in column seven (for rejections of the null hypothesis when negative estimates of the population variance occur) appears unnecessary; it provides little improvement in the test for randomness.

5. As the variation in the disturbances increases (relative to the variation due to the explanatory variable), the performance of the RCR estimator deteriorates. This is also true for the power of the test for significance of the coefficient means.

6. The relationship between the variability in the X variable and the performance of the estimators and test statistics is not clear. In the two cases examined here, $\sigma_X = 1.0$ and $\sigma_X = 10.0$, it would appear that the estimators and tests perform better when there is more variation in the explanatory variable (other things being equal). However, there may be a more complicated relationship operating here as Swamy (1971, pp. 38-48) found for the error components model, for example.

Note again that the RCR procedures perform best when both coefficients are random. When both coefficients are fixed the RCR model is inappropriate. An appropriate model in this case would be the classical pooling (CP) model of Chapter 2. For comparison purposes, results of applying the CP estimator and tests (allowing for different disturbance variances for each cross-section as in equations (2.6), (2.7), and (2.8)) are shown in Tables 8.12 through 8.21. The model specifications in these tables is exactly as it was in Tables 8.2 through 8.11, respectively, so direct comparisons of the performance of the estimators and tests can be made. The summary of the CP results is shown in the three columns of each panel of each table.

TABLE 8.12 Results of CP Estimation When $\beta_{0i} \sim N(0,1)$, $\beta_{1i} \sim N(0,1)$, $X_{it} \sim N(0,1)$ and $\varepsilon_{it} \sim N(0,\sigma_\varepsilon)$

		Estimated Coefficient	Percent Rejections $H_0: \bar{\beta}_k = 0$	Percent Coefficients Contained in 95% Confidence Interval
I. $\sigma_\epsilon = 1.0$	$\bar{\beta}_0$	0.004	72.2	29.3
	$\bar{\beta}_1$	0.005	72.1	29.7
II. $\sigma_\epsilon = 10.0$	$\bar{\beta}_0$	-0.001	63.7	37.9
	$\bar{\beta}_1$	-0.007	63.4	38.8
III. $\sigma_\epsilon = 20.0$	$\bar{\beta}_0$	-0.007	61.3	41.7
	$\bar{\beta}_1$	-0.010	57.7	45.6
IV. $\sigma_\epsilon \sim U(0,20)$	$\bar{\beta}_0$	-0.002	65.8	36.5
	$\bar{\beta}_1$	0.012	67.0	35.3
V. $\sigma_\epsilon \sim U(0, 40)$	$\bar{\beta}_0$	-0.007	62.8	39.6
	$\bar{\beta}_1$	0.015	64.2	39.2

Column one shows the estimated coefficient value. The assumption made is that the coefficients are fixed and equal for each cross-sectional unit. This is the correct assumption for the specification in Tables 8.20 and 8.21 but incorrect (at least partially) in the other tables.

Column two shows the percentage of times the hypothesis H_0: $\bar{\beta}_k = 0$ was rejected at a nominal 5% level of significance. For the CP model the test again assumes fixed and equal coeffi-

TABLE 8.13 Results of CP Estimation When $\beta_{0i} \sim N(0,1)$, $\beta_{1i} \sim N(0,1)$, $X_{it} \sim N(0,10)$ and $\varepsilon_{it} \sim N(0,\sigma_\varepsilon)$

		Estimated Coefficient	Percent Rejections $H_0:\beta_k = 0$	Percent Coefficients Contained in 95% Confidence Interval
I. $\sigma_\epsilon = 1.0$	$\bar{\beta}_0$	0.004	72.2	29.3
	$\bar{\beta}_1$	0.005	97.4	3.3
II. $\sigma_\epsilon = 10.0$	$\bar{\beta}_0$	-0.001	63.7	37.9
	$\bar{\beta}_1$	-0.008	97.1	3.2
III. $\sigma_\epsilon = 20.0$	$\bar{\beta}_0$	-0.007	61.3	41.7
	$\bar{\beta}_1$	-0.013	96.5	3.8
IV. $\sigma_\epsilon \sim U(0,20)$	$\bar{\beta}_0$	-0.002	65.8	36.5
	$\bar{\beta}_1$	0.010	96.8	3.3
V. $\sigma_\epsilon \sim U(0, 40)$	$\bar{\beta}_0$	-0.007	62.8	39.6
	$\bar{\beta}_1$	0.015	96.0	4.3

cients. Thus, the hypothesis test is to determine whether this common coefficient is equal to zero (rather than to determine whether the mean of possibly random coefficients is zero as in the RCR model).

The third column shows the percentage of times a 95% confidence interval contained the true value of the fixed coefficients or the mean of the random coefficients.

TABLE 8.14 Results of CP Estimation When $\beta_{0i} \sim N(1,1)$, $\beta_{1i} \sim N(1,1)$, $X_{it} \sim N(0,1)$ and $\varepsilon_{it} \sim N(0,\sigma_\varepsilon)$

		Estimated Coefficient	Percent Rejections $H_0:\beta_k = 0$	Percent Coefficients Contained in 95% Confidence Interval
I. $\sigma_\epsilon = 1.0$	$\bar{\beta}_0$	1.004	100.0	29.3
	$\bar{\beta}_1$	1.005	100.0	29.7
II. $\sigma_\epsilon = 10.0$	$\bar{\beta}_0$	0.999	99.9	37.9
	$\bar{\beta}_1$	0.993	99.6	38.8
III. $\sigma_\epsilon = 20.0$	$\bar{\beta}_0$	0.993	98.4	41.7
	$\bar{\beta}_1$	0.990	96.9	45.6
IV. $\sigma_\epsilon \sim U(0,20)$	$\bar{\beta}_0$	0.998	99.7	36.5
	$\bar{\beta}_1$	1.012	99.4	35.3
V. $\sigma_\epsilon \sim U(0, 40)$	$\bar{\beta}_0$	0.993	98.5	39.6
	$\bar{\beta}_1$	1.015	98.7	39.2

Examining Tables 8.12 through 8.21 the following observations for the CP estimators and test statistics can be made:

The CP estimator of the fixed coefficient performs well when the coefficient is fixed. This is also true for the fixed coefficient in the mixed models of Tables 8.18 and 8.19. The estimator averages and the enclosure rates for the confidence intervals both support this conclusion.

TABLE 8.15 Results of CP Estimation When $\beta_{0i} \sim N(1,1)$, $\beta_{1i} \sim N(1,1)$, $X_{it} \sim N(0,10)$ and $\varepsilon_{it} \sim N(0,\sigma_\varepsilon)$

		Estimated Coefficient	Percent Rejections $H_0: \beta_k = 0$	Percent Coefficients Contained in 95% Confidence Interval
I. $\sigma_\epsilon = 1.0$	$\bar{\beta}_0$	1.004	100.0	29.3
	$\bar{\beta}_1$	1.005	100.0	3.3
II. $\sigma_\epsilon = 10.0$	$\bar{\beta}_0$	0.999	99.9	37.9
	$\bar{\beta}_1$	0.992	100.0	3.2
III. $\sigma_\epsilon = 20.0$	$\bar{\beta}_0$	0.993	98.4	41.7
	$\bar{\beta}_1$	0.987	99.8	3.8
IV. $\sigma_\epsilon \sim U(0,20)$	$\bar{\beta}_0$	0.998	99.7	36.5
	$\bar{\beta}_1$	1.010	100.0	3.3
V. $\sigma_\epsilon \sim U(0, 40)$	$\bar{\beta}_0$	0.993	98.5	39.6
	$\bar{\beta}_1$	1.015	100.0	4.3

When coefficients are random, the CP estimator appears to be unbiased. A proof that this is the case can be constructed along lines similar to the proof by Zellner (1969) that the aggregation estimator will be unbiased when the coefficients are random. See Appendix A for the proof. The problem with using CP when coefficients are random is not the bias in the estimates but in the

TABLE 8.16 Results of CP Estimation When $\beta_{0i} \sim N(0,1)$, $\beta_{1i} \sim N(1,1)$, $X_{it} \sim N(0,1)$ and $\varepsilon_{it} \sim N(0,\sigma_\varepsilon)$

		Estimated Coefficient	Percent Rejections $H_0:\beta_k = 0$	Percent Coefficients Contained in 95% Confidence Interval
I. $\sigma_\epsilon = 1.0$	$\bar{\beta}_0$	0.004	72.2	29.3
	$\bar{\beta}_1$	1.005	100.0	29.7
II. $\sigma_\epsilon = 10.0$	$\bar{\beta}_0$	-0.001	63.7	37.9
	$\bar{\beta}_1$	0.993	99.6	38.8
III. $\sigma_\epsilon = 20.0$	$\bar{\beta}_0$	-0.007	61.3	41.7
	$\bar{\beta}_1$	0.990	96.9	45.6
IV. $\sigma_\epsilon \sim U(0,20)$	$\bar{\beta}_0$	-0.002	65.8	36.5
	$\bar{\beta}_1$	1.012	99.4	35.3
V. $\sigma_\epsilon \sim U(0, 40)$	$\bar{\beta}_0$	-0.007	62.8	39.6
	$\bar{\beta}_1$	1.015	98.7	39.2

performance of the hypothesis test for significance of the coefficients and in the performance of confidence interval estimators.

For example, comparing Tables 8.2 and 8.12, the RCR hypothesis test for significance is obviously superior to the CP test. The CP test has rejection rates much higher than the 5% level of significance set for the test. The RCR rejection rates are much closer to the nominal 5% level.

TABLE 8.17 Results of CP Estimation When $\beta_{0i} \sim N(0,1)$, $\beta_{1i} \sim N(1,1)$, $X_{it} \sim N(0,10)$ and $\varepsilon_{it} \sim N(0,\sigma_\varepsilon)$

		Estimated Coefficient	Percent Rejections $H_0: \beta_k = 0$	Percent Coefficients Contained in 95% Confidence Interval
I. $\sigma_\epsilon = 1.0$	$\bar{\beta}_0$	0.004	72.2	29.3
	$\bar{\beta}_1$	1.005	100.0	3.3
II. $\sigma_\epsilon = 10.0$	$\bar{\beta}_0$	-0.001	63.7	37.9
	$\bar{\beta}_1$	0.992	100.0	3.2
III. $\sigma_\epsilon = 20.0$	$\bar{\beta}_0$	-0.007	61.3	41.7
	$\bar{\beta}_1$	0.987	99.8	3.8
IV. $\sigma_\epsilon \sim U(0,20)$	$\bar{\beta}_0$	-0.002	65.8	36.5
	$\bar{\beta}_1$	1.010	100.0	3.3
V. $\sigma_\epsilon \sim U(0, 40)$	$\bar{\beta}_0$	-0.007	62.8	39.6
	$\bar{\beta}_1$	1.015	100.0	4.3

Also note that the CP enclosure rates for the 95% confidence intervals are very low when the coefficients are random.

In sum it would appear that the RCR and CP models perform well when the respective required assumptions are met. However, both deteriorate rapidly when used improperly. This suggests the importance of being able to choose the assumptions which are appropriate in each particular situation. The RCR test for

TABLE 8.18 Results of CP Estimation When $\beta_{0i} = 1.0$, $\beta_{1i} \sim N(1,1)$, $X_{it} \sim N(0,1)$ and $\varepsilon_{it} \sim N(0, \sigma_\varepsilon)$

		Estimated Coefficient	Percent Rejections $H_0: \beta_k = 0$	Percent Coefficients Contained in 95% Confidence Interval
I. $\sigma_\epsilon = 1.0$	$\bar{\beta}_0$	1.001	100.0	94.4
	$\bar{\beta}_1$	1.005	100.0	29.7
II. $\sigma_\epsilon = 10.0$	$\bar{\beta}_0$	0.997	100.0	93.7
	$\bar{\beta}_1$	0.993	99.6	38.8
III. $\sigma_\epsilon = 20.0$	$\bar{\beta}_0$	0.993	100.0	93.5
	$\bar{\beta}_1$	0.990	96.9	45.6
IV. $\sigma_\epsilon \sim U(0,20)$	$\bar{\beta}_0$	0.999	100.0	95.1
	$\bar{\beta}_1$	1.012	99.4	35.3
V. $\sigma_\epsilon \sim U(0, 40)$	$\bar{\beta}_0$	0.997	100.0	95.3
	$\bar{\beta}_1$	1.015	98.7	39.2

randomness should prove useful in this respect. As shown in the simulations, this test appears to perform well even in small samples in terms of being able to discriminate between fixed and random coefficients.

Finally, it was noted that the RCR estimators performed especially poorly in those cases when coefficients were fixed. Also,

TABLE 8.19 Results of CP Estimation When $\beta_{0i} = 1.0$, $\beta_{1i} \sim N(1,1)$, $X_{it} \sim N(0,10)$ and $\varepsilon_{it} \sim N(0,\sigma_\varepsilon)$

		Estimated Coefficient	Percent Rejections $H_0:\beta_k = 0$	Percent Coefficients Contained in 95% Confidence Interval
I. $\sigma_\epsilon = 1.0$	$\bar{\beta}_0$	1.001	100.0	94.4
	$\bar{\beta}_1$	1.005	100.0	3.3
II. $\sigma_\epsilon = 10.0$	$\bar{\beta}_0$	0.997	100.0	93.7
	$\bar{\beta}_1$	0.992	100.0	3.2
III. $\sigma_\epsilon = 20.0$	$\bar{\beta}_0$	0.993	100.0	93.5
	$\bar{\beta}_1$	0.987	99.8	3.8
IV. $\sigma_\epsilon \sim U(0,20)$	$\bar{\beta}_0$	0.999	100.0	95.1
	$\bar{\beta}_1$	1.010	100.0	3.3
V. $\sigma_\epsilon \sim U(0, 40)$	$\bar{\beta}_0$	0.997	100.0	95.3
	$\bar{\beta}_1$	1.015	100.0	4.3

in those cases, a number of negative estimates of the coefficient variances were obtained. Tables 8.22 through 8.25 provide some insight into the extent of the deterioration of RCR estimators due to the negative variance estimates.

Tables 8.22 and 8.23 show RCR estimation results when both coefficients are fixed and the X variable values were generated

TABLE 8.20 Results of CP Estimation When $\beta_{0i} = 1.0$, $\beta_{1i} = 1.0$, $X_{it} \sim N(0,1)$ and $\varepsilon_{it} \sim N(0, \sigma_\varepsilon)$

		Estimated Coefficient	Percent Rejections $H_0: \beta_k = 0$	Percent Coefficients Contained in 95% Confidence Interval
I. $\sigma_\epsilon = 1.0$	$\bar{\beta}_0$	1.001	100.0	94.4
	$\bar{\beta}_1$	1.001	100.0	94.6
II. $\sigma_\epsilon = 10.0$	$\bar{\beta}_0$	0.997	100.0	93.7
	$\bar{\beta}_1$	1.001	100.0	95.5
III. $\sigma_\epsilon = 20.0$	$\bar{\beta}_0$	0.993	100.0	93.5
	$\bar{\beta}_1$	1.003	100.0	94.6
IV. $\sigma_\epsilon \sim U(0,20)$	$\bar{\beta}_0$	0.999	100.0	95.1
	$\bar{\beta}_1$	1.002	100.0	95.6
V. $\sigma_\epsilon \sim U(0, 40)$	$\bar{\beta}_0$	0.997	100.0	95.3
	$\bar{\beta}_1$	1.001	100.0	95.9

as N(0,1). Tables 8.24 and 8.25 provide results under the same assumptions except that the X values were generated as N(0,10). In Tables 8.22 and 8.24, the results are summarized only in those cases when variances of both coefficients had positive estimates. In Tables 8.23 and 8.25, the results are summarized in cases when the estimate of variance of either or both coefficients was

TABLE 8.21 Results of CP Estimation When $\beta_{0i} = 1.0$, $\beta_{1i} = 1.0$, $X_{it} \sim N(0,10)$ and $\varepsilon_{it} \sim N(0, \sigma_\varepsilon)$

		Estimated Coefficient	Percent Rejections $H_0: \beta_k = 0$	Percent Coefficients Contained in 95% Confidence Interval
I. $\sigma_\epsilon = 1.0$	$\bar{\beta}_0$	1.001	100.0	94.4
	$\bar{\beta}_1$	1.001	100.0	94.6
II. $\sigma_\epsilon = 10.0$	$\bar{\beta}_0$	0.997	100.0	93.7
	$\bar{\beta}_1$	1.000	100.0	95.5
III. $\sigma_\epsilon = 20.0$	$\bar{\beta}_0$	0.993	100.0	93.5
	$\bar{\beta}_1$	1.000	100.0	94.6
IV. $\sigma_\epsilon \sim U(0,20)$	$\bar{\beta}_0$	0.999	100.0	95.1
	$\bar{\beta}_1$	1.000	100.0	95.6
V. $\sigma_\epsilon \sim U(0, 40)$	$\bar{\beta}_0$	0.997	100.0	95.3
	$\bar{\beta}_1$	1.000	100.0	95.9

negative. The number of cases where positive or negative variances occurred is given on the left hand side of each table under the disturbance standard deviation values. Note the high number of negative variance estimates in these cases.

The tables show the estimate of the coefficient mean (fixed coefficient in this case) in column one and the standard deviation

TABLE 8.22 Results of RCR Estimation When $\beta_{0i} = 1.0$, $\beta_{1i} = 1.0$, $X_{it} \sim N(0,1)$ and $\varepsilon_{it} \sim N(0,\sigma_\varepsilon)$ and Positive Estimates of Coefficient Variances Resulted

		Estimated Mean	Standard Deviation of Estimated Mean
I. $\sigma_\epsilon = 1.0$ (N = 220)	$\bar{\beta}_0$	1.002	0.053
	$\bar{\beta}_1$	0.999	0.059
II. $\sigma_\epsilon = 10.0$ (N = 98)	$\bar{\beta}_0$	1.016	0.244
	$\bar{\beta}_1$	0.972	0.422
III. $\sigma_\epsilon = 20.0$ (N = 84)	$\bar{\beta}_0$	1.120	0.530
	$\bar{\beta}_1$	1.000	0.713
IV. $\sigma_\epsilon \sim U(0, 20)$ (N = 129)	$\bar{\beta}_0$	0.971	0.319
	$\bar{\beta}_1$	0.982	0.333
V. $\sigma_\epsilon \sim U(0, 40)$ (N = 108)	$\bar{\beta}_0$	1.310	1.388
	$\bar{\beta}_1$	1.895	1.197

or standard error of this estimate in column two. The standard errors are computed as the square root of the diagonal elements of the variance-covariance matrix in equation (5.41).

The tables clearly indicate that negative variance estimates create problems for the mean estimator and the precision of the estimates as measured by the standard error. When both coefficient

TABLE 8.23 Results of RCR Estimation When $\beta_{0i} = 1.0$, $\beta_{1i} = 1.0$, $X_{it} \sim N(0,1)$ and $\varepsilon_{it} \sim N(0, \sigma_\varepsilon)$ and Negative Estimates of Coefficient Variances Resulted

		Estimated Mean	Standard Deviation of Estimated Mean
I. $\sigma_\epsilon = 1.0$ (N = 780)	$\bar{\beta}_0$	0.980	0.337
	$\bar{\beta}_1$	0.986	0.346
II. $\sigma_\epsilon = 10.0$ (N = 902)	$\bar{\beta}_0$	0.881	3.507
	$\bar{\beta}_1$	0.945	3.361
III. $\sigma_\epsilon = 20.0$ (N = 916)	$\bar{\beta}_0$	1.301	11.279
	$\bar{\beta}_1$	4.829	113.832
IV. $\sigma_\epsilon \sim U(0,20)$ (N = 871)	$\bar{\beta}_0$	0.355	16.710
	$\bar{\beta}_1$	1.709	19.554
V. $\sigma_\epsilon \sim U(0, 40)$ (N = 892)	$\bar{\beta}_0$	1.123	1.524
	$\bar{\beta}_1$	0.987	4.604

variance estimates are positive the RCR estimator performs fairly well. When one or both of the variance estimates is negative, however, the deterioration of the estimates can be striking. Compare, for example, the results from Table 8.22, panel III and Table 8.23, panel III.

TABLE 8.24 Results of RCR Estimation When $\beta_{0i} = 1.0$, $\beta_{1i} = 1.0$, $X_{it} \sim N(0,10)$ and $\varepsilon_{it} \sim N(0, \sigma_\varepsilon)$ and Positive Estimates of Coefficient Variances Resulted

		Estimated Mean	Standard Deviation of Estimated Mean
I. $\sigma_\epsilon = 1.0$ (N = 220)	$\bar{\beta}_0$	1.002	0.053
	$\bar{\beta}_1$	1.000	0.006
II. $\sigma_\epsilon = 10.0$ (N = 98)	$\bar{\beta}_0$	1.016	0.244
	$\bar{\beta}_1$	0.997	0.042
III. $\sigma_\epsilon = 20.0$ (N = 84)	$\bar{\beta}_0$	1.122	0.534
	$\bar{\beta}_1$	1.000	0.073
IV. $\sigma_\epsilon \sim U(0,20)$ (N = 129)	$\bar{\beta}_0$	0.971	0.319
	$\bar{\beta}_1$	0.998	0.033
V. $\sigma_\epsilon \sim U(0, 40)$ (N = 108)	$\bar{\beta}_0$	1.310	1.389
	$\bar{\beta}_1$	0.990	0.120

These results further emphasize the need to correctly choose between random and fixed specifications. A correct choice will result in the absence of most negative variance estimates. When a coefficient is judged to be fixed, the appropriate components in the matrix $\hat{\Delta}$ can be set equal to zero (see Chapter 5, the discussion of the alternative estimators $\hat{\Delta}_1$, $\hat{\Delta}_2$, or $\hat{\Delta}_3$). Alternatively,

TABLE 8.25 Results of RCR Estimation When $\beta_{0i} = 1.0$, $\beta_{1i} = 1.0$, $X_{it} \sim N(0,10)$ and $\varepsilon_{it} \sim N(0,\sigma_\varepsilon)$ and Negative Estimates of Coefficient Variances Resulted

		Estimated Mean	Standard Deviation of Estimated Mean
I. $\sigma_\epsilon = 1.0$ (N = 780)	$\bar{\beta}_0$	0.978	0.355
	$\bar{\beta}_1$	0.999	0.040
II. $\sigma_\epsilon = 10.0$ (N = 902)	$\bar{\beta}_0$	0.853	4.210
	$\bar{\beta}_1$	0.992	0.374
III. $\sigma_\epsilon = 20.0$ (N = 916)	$\bar{\beta}_0$	1.663	21.936
	$\bar{\beta}_1$	1.921	22.961
IV. $\sigma_\epsilon \sim U(0,20)$ (N = 871)	$\bar{\beta}_0$	-4.693	121.689
	$\bar{\beta}_1$	1.489	14.227
V. $\sigma_\epsilon \sim U(0, 40)$ (N = 892)	$\bar{\beta}_0$	1.123	1.527
	$\bar{\beta}_1$	0.995	0.467

an estimator of Δ which did not allow negative variance estimates could be used (such as $\hat{\Delta}_2$ or $\hat{\Delta}_3$ mentioned above). Which estimator of Δ performs "best" is a matter of empirical investigation.

In this Monte Carlo experiment no examination of mixed RCR techniques (use of $\hat{\Delta}_1$) has been made. It would appear that use of the mixed RCR model would combine the best aspects of RCR and CP models and alleviate certain of the problems encountered by these estimators. Further simulation studies should make a point of examining results from the mixed model. Even though this simulation has been limited in scope (as all simulations must be) it is hoped that it will shed some light on performance of RCR estimators and tests in small samples.

8.3 SOME SUGGESTIONS FOR CHOOSING AN APPROPRIATE MODEL

A researcher familiar with his or her particular discipline but inexperienced in dealing with time series observations on a number of cross-sectional units may find the task of choosing a particular method of analysis somewhat confusing. As pooled data bases become increasingly common the need for a more structured approach to the choice of a model is evident. In this section an attempt is made at the provision of such structure through the presentation of guidelines for the analysis of pooled cross-sectional and time series data. These guidelines will involve a number of basic questions the researcher will need to answer. The answers to these questions should help lead to an appropriate method for analysis. Figure 8.1, which partially illustrates the choice process, will be referred to throughout this section.

In choosing the appropriate methodology the researcher must first ask whether coefficients for each individual unit are equal or whether they differ.

Hypothesis tests are available to aid in making this decision. In Chapter 2, for example, the test statistic in equation (2.30) can be used for testing whether all or a subset of the coefficients in the model are equal. (Also, see Maddala (1977, pp. 322-336.)) An alternative test procedure is examined by Toro-Vizcarrondo and Wallace (1968), Wallace and Toro-Vizcarrondo (1969), Goodnight and Wallace (1972), Wallace (1972) and McElroy (1977b).

Wallace (1972) shows that, even though coefficients may differ between individual units, the variance obtained from estimating them as equal is smaller than the individual estimator variances. But the pooled estimates will be biased. Wallace suggests that a tradeoff, accepting some bias in order to reduce variances, may be acceptable in some cases. Accordingly, the test is based on mean square error and includes tables of critical values in Goodnight and Wallace (1972). The test is used to compare whether CP or separate regressions would be preferable.

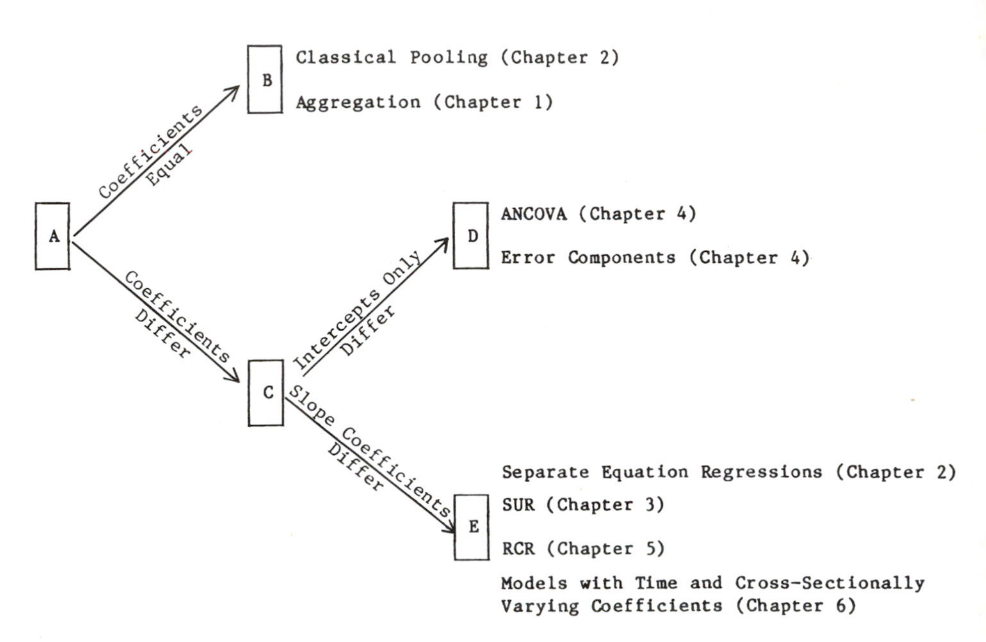

FIGURE 8.1 Guidelines for model choice.

Zellner (1962) developed a test for equality of coefficients when contemporaneous correlation is present. The test statistic is shown in equation (3.17). Swamy's (1970) test statistic for equality of coefficients is shown in equation (5.55). This statistic is computed assuming no contemporaneous correlation and is a special case of Zellner's statistic. Both the Swamy (1970) and Zellner (1962) tests allow the disturbance variance to differ between cross-sectional units.

Suppose that coefficients are judged to be equal. Then the upper branch of the diagram to node B is chosen and suitable methods for analysis would be either aggregation (Chapter 1) or classical pooling (Chapter 2).

On the other hand, if coefficients are judged to differ then the lower branch from node A to node C is selected. At this point the researcher must decide whether the intercepts differ and slope coefficients are fixed or whether slope coefficients differ also.

If the intercepts differ while the slope coefficients do not, the branch from node C to node D is selected. As can be seen, the ANCOVA and error components models are suitable methods for analysis. ANCOVA provides estimates of the change in cross-section intercepts with respect to a base level individual unit if such estimates are desired. The desire for these estimates occurs more frequently in the design of experiments where the "covariates" or variables in the regression equation are used to adjust the estimates of the intercepts. It is the intercepts in which primary interest is centered, however. In many applications the intercept estimates provide minimal information and there is little interest in the actual values. The error components model would be useful if interest centered on the coefficients of the explanatory variables rather than the intercepts since it treats the intercept as a random variable with the value for an individual viewed as being drawn from a population described by a mean and variance. The mean and variance are then estimated and used to adjust the coefficient estimates. Since the number of parameters to be estimated is considerably reduced by the error components approach, its use should be considered when the intercepts add little or no information to an analysis. Also, as pointed out by Jakubson and Kiefer (1985) in a comment on an article by Hsiao (1985), the error components approach would be applicable when forecasts are desired since the distribution of the random effect provides necessary information for developing the forecasts.

In choosing between the ANCOVA and EC models the choice is one of whether to treat the intercepts as fixed or random. Mundlak (1978a) suggests that the intercepts can always be considered random. When the ANCOVA estimator is used the inferences are simply conditional on the intercepts observed in the sample. Inferences from the EC model are unconditional but assumptions about the distribution of the intercepts must be imposed which are unnecessary when ANCOVA is used.

A further consideration noted by Mundlak (1978a) Chamberlain (1978, 1984), Hausman and Taylor (1981) and others is whether the intercepts are correlated with the independent variables. If so, the GLS estimator will be biased but the ANCOVA estimator will be unbiased, conditional on the intercepts observed in the sample. As noted by Judge, et al. (1985, p. 550) one alternative in this case is to use EC if the intercepts can be presumed independent of the explanatory variables and use ANCOVA if correlation is present. Tests have been developed to help determine if the random effects and the independent variables are correlated. These tests are discussed in Chapter 4 and are due to Hausman (1978) and Chamberlain (1982, 1984). Also useful in the error components model are tests to determine whether the intercepts vary over time and/or individuals. The test procedure developed by Breusch and Pagan (1980) for this purpose is also discussed in Chapter 4.

Small sample results in Section 4.5 also play a role in the choice between EC and ANCOVA. Swamy and Arora (1972) suggest that N − K (and T − K if time effects are present), where K is the number of explanatory variables, both be larger than ten before EC estimators are used. Taylor (1980) arrives at the same conclusion. If N and T are small and the intercept variance is large the ANCOVA estimator should be used. If N and T are small and there is little variation in the intercepts, use the CP estimator.

If the coefficients of the explanatory variables are thought to vary either over time or individuals then the branch of the diagram from node C to node E is followed. In this case the intercepts can vary as well as the structural coefficients. Some of the tests available to assess variation in the coefficients of the explanatory variables have already been mentioned. These include the test statistic in equation (2.30) of Chapter 2 (and the alternate procedure based on the mean square error criterion), the Zellner (1962) test in equation (3.17) and Swamy's test in equation (5.55). Which of these tests is appropriate depends on assumptions made concerning disturbances. The test statistic in equation (2.30) assumes equal disturbance variances for each cross-section. Swamy's test statisic allows variances to differ and Zellner's test also assumes disturbances may be contemporaneously correlated. These tests are all designed to determine whether the regression coefficients vary across individuals.

In Chapter 6 tests were discussed to determine whether coefficients varied randomly over time. Depending on whether the Hsiao model or the Swamy and Mehta model is used, appropriate test statistics are discussed in Chapter 6 for this test.

Suppose that slope coefficients were judged to differ and the branch from node C to node E is taken. Also suppose that coefficients vary over individuals but not over time.

There are several alternative procedures available at this point. If coefficients are assumed to be fixed but different, separate equation regressions or SUR (Chapter 3) would be appropriate methods of analysis. However, if coefficents are assumed to vary randomly between cross-sectional units, Swamy's RCR model (Chapter 5) may be appropriate. There is no direct test to distinguish between the fixed-but-different and random specifications. The researcher must, therefore, base this choice on other considerations.

If interest lies in estimating a population average coefficient and either separate equation regressions or SUR is used then a simple average of the coefficients or some type of weighted average must be constructed. On the other hand, the RCR methodology computes a weighted average of the coefficient estimates and also provides well-based procedures for inferences about population means and variances. Thus the RCR procedures provide a more direct pooled analysis.

Mundlak and Yahav (1981) considered the question of whether coefficients should be treated as fixed-but-different or random. They suggest, as with the error components model, that the coefficients can always be considered random. If separate equation regressions or SUR is used the inferences would simply be considered conditional on the values of the coefficients that actually occur in the sample. Of course, if Swamy's RCR model is used, more efficient estimates will be obtained when the RCR assumptions are reasonable.

One instance when the RCR assumptions are violated is when the random coefficients and the explanatory variables are correlated. This problem is discussed by Mundlak and Yahav (1981), Chamberlain (1982) and Swamy (1971, pp. 143). Swamy suggested using mixed RCR (Chapter 5) procedures as one way to avoid this problem. Treat the coefficients as fixed but different when correlation is suspected and as random otherwise.

Judge et al. (1985, pp. 544-545) suggest using a test developed by Pudney (1978) to test for correlation between random coefficients and explanatory variables.

The number of individuals in the sample is also a consideration in choosing whether to use the RCR procedures. The sample size, N, should be sufficiently large to provide good estimates of the coefficient variances. The guidelines in the RCR case are not as well defined as in the EC model. How large the sample size must be in order to provide reliable variance estimates is a question requiring further research. The simulation results in the previous section of this chapter suggest that N = 20 is large enough to expect the RCR estimators to perform well, provided the other model assumptions are not violated. The results of Carter and Yang (1986) also suggest that T need not be large for the RCR procedures to be applied as long as N is large. Again, the question of how large N and/or T must be to provide reliable estimates and inferences is an empirical one.

If coefficients vary over time as well as cross-sectionally, the time and cross-sectionally varying coefficient models of Chapter 6 are appropriate.

This discussion is not intended to be the final word in analyzing pooled cross-sectional and time series data. The individual researcher may want to ammend the chart according to the needs of his or her own discipline. Also, the chart may be altered as research continues on new methods of analysis. The flowchart was purposefully kept brief since its aim is to provide the empirical researcher only with guidelines concerning the basic techniques for analyzing pooled cross-sectional and time series data and to indicate when the use of each of the techniques is appropriate. Also, the researcher should not feel restricted by the guidelines to the use of only one of the methodologies suggested in the flowchart. As Mehta, Narasimham, and Swamy (1978) suggested, it will often be informative to examine estimates from two or more of the techniques discussed and consider which estimates appear most reasonable given the researcher's prior knowledge of how variables should interact.

For other suggestions regarding choice of methodology see Dielman (1980, pp. 65-76) and Judge et al. (1985, p. 517, 550-551).

Finally, as always, some caution must be expressed regarding use of pretesting procedures to search for a model. See Judge and Bock (1978) concerning pretest estimators.

9
Suggestions for Further Research

9.1 VARIABLE SELECTION PROCEDURES

In separate equation regression models where least squares assumptions are satisfied, the effects of model misspecification (by omission of important variables) are widely known:

1. The coefficient estimates in the misspecified model will be biased.
2. The estimator in the misspecified model will have generally smaller variance than the estimator of the same coefficients in a properly specified "true" model.
3. There are certain conditions under which the gain in precision is not offset by the bias.

These conditions are explicity defined in Hocking (1976, pp. 5-7).

Results analogous to these for the general linear regression model will be presented in this section. These results are stated explicity here since they seem to be much less widely known.

The "true" model is written as

$$Y = X\beta + \varepsilon \tag{9.1}$$

$$= X_P \beta_P + X_R \beta_R + \varepsilon \tag{9.2}$$

where X has been partitioned into X_P and X_R, both of full rank, and β, the vector of coefficients to be estimated, has been partitioned into β_P and β_R. Y is a vector of observations on the dependent variable and ε is a random disturbance with

$$E(\varepsilon) = 0 \tag{9.3}$$

$$E(\varepsilon\varepsilon') = \Omega \tag{9.4}$$

where Ω is a symmetric positive definite (pd) matrix. The term "true" model suggests a model where all important variables have been included, that is, a properly specified model.

The generalized least squares estimator of β is

$$\tilde{\beta} = \begin{bmatrix} \tilde{\beta}_P \\ \tilde{\beta}_R \end{bmatrix} = (X'\Omega^{-1}X)^{-1}X'\Omega^{-1}Y \tag{9.5}$$

The misspecified model is

$$Y = X_P\beta_P + \varepsilon \tag{9.6}$$

and the GLS estimator of β_P is

$$\tilde{\beta}_P^* = (X_P'\Omega^{-1}X_P)^{-1}X_P'\Omega^{-1}Y \tag{9.7}$$

Rosenberg and Levy (1972) summarize the main results:

1. $\tilde{\beta}_P^*$ will be biased unless $\beta_R = 0$.
2. Var $(\tilde{\beta}_P)$ − Var $(\tilde{\beta}_P^*)$ is <u>positive semidefinite</u> (psd). Thus, the estimates of β_P from the "true" model are generally more variable than those from the misspecified model.

It is also easy to show that the following result holds:

3. If Var $(\tilde{\beta}_R) - \beta_R\beta_R'$ is psd, then so is Var$(\tilde{\beta}_P)$ − MSE $(\tilde{\beta}_P^*)$ where MSE$(\tilde{\beta}_P^*)$ represents the mean square error of the estimator of β_P in the misspecified model. This result describes a condition under which the gain in precision in the misspecified model is not offset by the bias. For proofs of the above results, see Appendix B.

Hocking's result can now be easily extended to many of the models considered in ths text. As an example, consider the RCR model of Chapter 5.

For the ith individual unit in our sample of N cross-sectional units the time series regression model is specified as in equation (5.1):

$$Y_i = X_i\beta_i + \varepsilon_i \tag{9.8}$$

Assume that assumptions 5.1 to 5.5 hold.

The equations given by (9.8) can be written as:

$$Y = X\bar{\beta} + e \tag{9.9}$$

where Y, X, $\bar{\beta}$, and e are defined in equations (5.5). It follows that $E(ee') = \Omega$ as shown in equation (5.6).

To examine the effect of misspecification on RCR estimates of the coefficient means, partition X in equation (9.9) as

$$X = \begin{bmatrix} X_{1P} & X_{1R} \\ X_{2P} & X_{2R} \\ \cdot & \cdot \\ \cdot & \cdot \\ X_{NP} & X_{NR} \end{bmatrix} = \begin{bmatrix} X_P & X_R \end{bmatrix} \tag{9.10}$$

and $\bar{\beta}$ as

$$\bar{\beta} = \begin{bmatrix} \bar{\beta}_P \\ \bar{\beta}_R \end{bmatrix} \tag{9.11}$$

The "true" model is then

$$Y = X_P\bar{\beta}_P + X_R\bar{\beta}_R + e \tag{9.12}$$

while the misspecified model is

$$Y = X_P\bar{\beta}_P + e \tag{9.13}$$

The GLS estimate of $\bar{\beta}$ in the true model is

$$\tilde{\bar{\beta}} = \begin{bmatrix} \tilde{\bar{\beta}}_P \\ \tilde{\bar{\beta}}_R \end{bmatrix} = (X'\Omega^{-1}X)^{-1}X'\Omega^{-1}Y \tag{9.14}$$

and for $\bar{\beta}_P$ in the misspecified model is

$$\tilde{\tilde{\beta}}_P^* = (X_P'\Omega^{-1}X_P)^{-1}X_P'\Omega^{-1}Y \qquad (9.15)$$

Applying the general linear model results to $\tilde{\tilde{\beta}}_P$ and $\tilde{\tilde{\beta}}_P^*$:

1. $\tilde{\tilde{\beta}}_P^*$ will be biased unless $\bar{\beta}_R = 0$.
2. Var ($\tilde{\tilde{\beta}}_P$) − Var ($\tilde{\tilde{\beta}}_P^*$) is psd.
3. If Var ($\tilde{\tilde{\beta}}_R$) − $\bar{\beta}_R\bar{\beta}_R'$ is psd then so is Var ($\tilde{\tilde{\beta}}_P$) − MSE ($\tilde{\tilde{\beta}}_P^*$).

Since $\tilde{\tilde{\beta}}$ is an unbiased estimator, Var ($\tilde{\tilde{\beta}}_P$) is equal to MSE ($\tilde{\tilde{\beta}}_P^*$). Thus, condition (3) indicates when $\tilde{\tilde{\beta}}_P^*$ will be superior to $\tilde{\tilde{\beta}}_P$ on the basis of a minimum mean square error criterion.

As in the OLS models analyzed by Hocking, there is a trade-off in the RCR model in choosing which variable subset is "best"; deleting variables with a nonzero mean can reduce the variance of the estimates of the remaining coefficients but it will also result in a bias in these estimates.

Hocking uses the results of his paper to suggest several selection criteria in deciding whether or not to delete variables in choosing an appropriate subset regression. From the results in this section for the general linear model it seems feasible to extend these criteria to more general regression models, for example, Swamy's RCR model. To determine whether parameter estimates might be obtained with smaller mean square error by inclusion or omission of certain variables would seem an important concept in random coefficient models as well as other models for pooled data. Questions to be examined might include

1. Are C_P–Plots (see Hocking (1976, pp. 16-19) useful in more general regression models?

2. How do results on forecasting as presented by Hocking apply in more general models?

3. Can estimation techniques suggested by Swamy (1973), including Stein-like estimators, ridge estimators and a minimum conditional mean square error estimator, be viewed as variable selection procedures in RCR models in the sense that Hocking views ridge regression in his paper?

9.2 TESTS FOR ASSUMPTIONS MADE ABOUT MODEL DISTURBANCES

In separate equation regression models there are tests for a variety of assumptions made concerning the disturbances. For example, there are tests for serial correlation and for heteroskedasticity. In pooled models these tests are not as well developed. One could apply the standard tests, such as the Durbin-Watson test for serial correlation, to the time series regression for each cross-sectional unit and treat each cross-sectional unit separately. But, considering the nature of the data, it would seem to make more sense to approach the problem as what might be called a global one. Rather than treating each cross-sectional unit separately, can a test be developed to detect the presence of serial correlation in all N cross-sectional units? This would seem more consistent with the concept of estimating the parameters of a model for all N cross-sectional units.

As an example consider again the RCR model of Chapter 5:

$$Y_i = X_i\beta_i + \varepsilon_i \tag{9.16}$$

for i = 1, ⋯,N as in equation (5.1). The terms are defined in equation (1.8).

If first order serial correlation of the disturbances exists this relationship can be written as:

$$\varepsilon_{it} = \rho_i\varepsilon_{i,t-1} + u_{it} \qquad i = 1, \cdots, N;\ t = 1, \cdots, T \tag{9.17}$$

where ε_{it} is the disturbance for the ith cross-sectional unit in time period t, $\varepsilon_{i,t-1}$ is the disturbance lagged one time period, ρ_i is the serial correlation coefficient for security i, and the u_{it} are independent disturbances with u_{it} being $N(0,\sigma_i^2)$.

The usual test performed in regression analysis tests, for each cross-sectional unit, the hypothesis

$$H_o\colon \rho_i = 0$$

versus

$$H_a\colon \rho_i \neq 0 \tag{9.18}$$

by using the Durbin-Watson statistic as in Johnston (1972, pp. 251-252) for example. Since concern here is with pooled cross-sectional and time series data, a global or pooled test for serial correlation might be of interest. The term pooled test means a test of the hypothesis

$$H_o\colon \rho_1 = \rho_2 = \cdots = \rho_N = \rho = 0$$

versus

$$H_a\text{: not all } \rho_i \text{ are equal to zero} \tag{9.19}$$

with level of significance α. This is not the same as testing the hypothesis in (9.18) with level of significance α for each of the N units. The hypothesis in (9.19) represents a simultaneous test of whether all serial correlation coefficients are zero.

To construct such a test the relationships in (9.17) can be written in vector notation as

$$\varepsilon_i = \rho_i \varepsilon_i^* + u_i \tag{9.20}$$

where

$$\varepsilon_i = \begin{bmatrix} \varepsilon_{i1} \\ \varepsilon_{i2} \\ \cdot \\ \cdot \\ \cdot \\ \varepsilon_{iT} \end{bmatrix} \quad \varepsilon_i^* = \begin{bmatrix} \varepsilon_{i0} \\ \varepsilon_{i1} \\ \cdot \\ \cdot \\ \cdot \\ \varepsilon_{i,T-1} \end{bmatrix} \quad \text{and } u_i = \begin{bmatrix} u_{i1} \\ u_{i2} \\ \cdot \\ \cdot \\ \cdot \\ u_{iT} \end{bmatrix} \tag{9.21}$$

Assume the u_i are $N(0, \sigma_i^2 I_T)$ and the ε_i^* and u_i are independent. If the ε_{it} values are assumed known, and if for each cross-sectional unit there are a large number of these values, then the N equations given by (9.20) can be viewed as a set of regression equations and the Chow test for coefficient equality or an asymptotically equivalent test developed by Zellner (1962) and extended by Swamy (1971, pp. 124-126) can be applied to test the hypothesis in (9.19). The Zellner/Swamy test allows variances to differ between individual time series and is also specifically designed for use in large samples. (See equation (5.55) and the preceding material for a discussion of the test.) The test statistic in such a case is

$$R = \sum_{i=1}^{N} \frac{\hat{\rho}_i \varepsilon_i'^* \varepsilon_i^* \hat{\rho}_i}{\hat{\sigma}_i^2} \tag{9.22}$$

$$= \sum_{i=1}^{N} \left[\frac{\hat{\rho}_i^2}{\hat{\sigma}_i^2} \cdot \sum_{t=1}^{T} \varepsilon_{i,t-1}^2 \right] \tag{9.23}$$

where

$$\hat{\rho}_i = (\varepsilon_i'^{*} \varepsilon_i^{*})^{-1} \varepsilon_i'^{*} \varepsilon_i \tag{9.24}$$

$$= \left(\sum_{t=1}^{T} \varepsilon_{i,t-1}^2 \right)^{-1} \sum_{t=1}^{T} \varepsilon_{i,t-1}\, \varepsilon_{it} \tag{9.25}$$

is the separate equation OLS estimate of the ith autocorrelation coefficient,

$$\hat{\sigma}_i^2 = \frac{\hat{u}_i'\hat{u}_i}{T - 1} \tag{9.26}$$

is the mean square error from the ith separate equation regression of ε_i and ε_i^*, and

$$\hat{u}_i = \varepsilon_i - \hat{\rho}_i \varepsilon_i^* \tag{9.27}$$

is the estimated residual vector from such a regression.

Under H_0, R has been shown to be asymptotically distributed as χ^2 with N degrees of freedom.

If the null hypothesis is accepted, conclude that first-order serial correlation of residuals is not present in any of the time series regressions. If the null hypothesis is rejected, conclude that serial correlation exists for at least one of the time series regressions.

Alternatively the following hypothesis can be tested:

$$H_o: \rho = 0$$

versus

$$H_a: \rho \neq 0 \tag{9.28}$$

where ρ is a common serial correlation coefficient. That is, assuming all the ρ_i in the model are equal to some common value, ρ, is this value equal to zero? The estimate of ρ is

$$\hat{\rho} = \left[\sum_{i=1}^{N} \frac{1}{s_i^2} (\varepsilon_i'^{*} \varepsilon_i^{*}) \right]^{-1} \sum_{i=1}^{N} \frac{1}{s_i^2} \varepsilon_i'^{*} \varepsilon_i \tag{9.29}$$

where s_i^2 is the estimate of σ_i^2. The test statistic,

$$t = \frac{\hat{\rho}}{se(\hat{\rho})} \tag{9.30}$$

where $se(\hat{\rho})$ is the standard error of the estimate, has a t-distribution with $N - 1$ degrees of freedom under H_0.

One approach taken to remedy the problem of serial correlation with separate equation OLS regressions is to estimate the ρ_i and use this estimate, $\hat{\rho}_i$, to transform the data in order to produce a series with noncorrelated residuals. Then re-estimate using the new series. This procedure is well known and is outlined in most econometric texts (see Pindyck and Rubinfeld, 1976, pp. 108-111 or Theil, 1971, pp. 250-254). For a pooled analysis it is possible to proceed in a similar fashion: transform each equation using the separately estimated $\hat{\rho}_i$ and then perform the pooled analysis. (See Swamy, 1971, pp. 127-132 or Parks, 1967, for examples.)

The problem encountered in using the test statistics presented in equation (9.22) or (9.30) is that the true residuals, the ε_{it}, are unknown (and unobservable). Therefore, in order to apply the test in practice, the ε_{it} must be estimated. The OLS residuals provide suitable estimates. These can be obtained by regressing Y_i on X_i for each of the N cross-sectional units. Performing these N time series regressions the estimated residuals are computed as

$$\hat{\varepsilon}_{it} = Y_i - \hat{\beta}_i X_i \tag{9.31}$$

where $\hat{\beta}_i$ is the OLS estimate of β_i.

This approach is justified asymptotically. Theil (1971, pp. 378-379) has shown that $\hat{\varepsilon}_{it} - \varepsilon_{it}$ converges in probability to zero as T approaches infinity. Therefore in large samples the estimated residual, $\hat{\varepsilon}_{it}$, should provide a good approximation of the true disturbance, ε_{it}. The effectiveness of tests based on such asymptotic theory remains to be seen when applied in small samples, however.

The $\hat{\varepsilon}_{it}$ can be used to estimate the ρ_i by substituting the $\hat{\varepsilon}_{it}$ for the ε_{it} in equation (9.21). Then perform the regressions

$$\hat{\varepsilon}_i = \rho_i \hat{\varepsilon}_i^* + u_i \tag{9.32}$$

As Johnston (1972, pp. 303-307) points out, if the u_{it}'s are normally and independently distributed, least squares applied to (9.32) will yield consistent estimates of the ρ_i. The presence of the lagged explanatory variable will cause no problems in terms of large sample properties. In small samples the ρ_i will be biased where the bias is of order less than 1/T.

Using the $\hat{\rho}_i$ obtained from the regressions performed in (9. 32) the test statistic R can be constructed to determine whether there is significant serial correlation. The convergence property insures that R calculated using the $\hat{\varepsilon}_{it}$ will have the same distribution asymptotically as if the true ε_{it} were used.

Other tests could possibly be developed in a manner similar to the test for serial correlation. These might include a test for heteroskedasticity and tests for higher order serial correlation.

The test for serial correlation is justified asymptotically but its behavior in small samples is unknown. Monte Carlo simulation could be used to examine whether this test, as well as others, performed well in small samples. This would also be true for other tests discussed in previous chapters.

Further research might also concentrate on the effect of violations of assumptions on the parameter estimators in pooled models. Again, asymptotically, it is well known that violations of certain assumptions will result in less efficient estimators if the violations are not detected and some adjustment made. This is true when distrubances are serially correlated, for example. However, it is unclear how severe serial correlation must be to cause a significant deterioration in the efficiency of estimators. Simulation studies might shed some light in this area. The effect of violations of assumptions on inferences from pooled models would be another area of interest.

As an example of the effects of serial corrrelation in a pooled model consider a study by Dielman, Nantell, and Wright (1980). The study involved 70 firms which announced a repurchase of their own stock by way of tender offers. For each of these 70 firms monthly rates of return for the securities were obtained from the Center for Research on Security Prices (CRSP) tapes. Returns for 107 months preceding the repurchase announcement and 60 months following the announcement were included. The return on the market was also taken from the CRSP data set. As a risk free rate the monthly rates on three-month treasury bills were used. These rates are compiled by the Federal Reserve and were obtained from the NBER (National Bureau of Economic Research) data base.

Initially, consider the following models:

$$R_{it} - R_{Ft} = \beta_{oi} + \beta_{li}(R_{mt} - R_{Ft}) + \varepsilon_{it} \quad (9.33)$$

where

R_{it} = the return for security i in month t.

R_{mt} = the return on the market in month t.

R_{Ft} = the risk free rate in month t.

β_{oi} and β_{li} are regression parameters to be estimated for the ith security, and ε_{it} is a disturbance term.

Random coefficient regression (RCR) was used to estimate the population average parameters $\bar{\beta}_o$ and $\bar{\beta}_l$, using the technique outlined in Chapter 5.

Table 9.1 provides the initial RCR estimates for the model in (9.33) with both coefficients assumed random. The estimate of the average intercept is 0.0017; the estimate of the mean of the beta coefficients (the slope coefficients) is 1.2093. Note that the estimate of the population variance of the intercept was negative. This indicates either

1. a model misspecification occurs in terms of the error covariance structure.
2. the coefficient is nonrandom.

To investigate the error covariance structure of the model the pooled test for serial correlation was applied and the results are summarized in Table 9.2.

The average correlation coefficient for the 70 firms was estimated as $\hat{\rho} = -0.0858$. This average is a weighted average of the OLS estimates of the individual firm autocorrelation coefficients from the regressions shown in equation (9.33). Each estimate is weighted by its precision and these products are added together to obtain $\hat{\rho}$.

The test results indicate that the serial correlation coefficients are not all equal to zero. The χ^2 value of 166.6 is above the 1% critical value, 100.5. Thus the test suggests that there is significant serial correlation for at least one security.

The presence of serial correlation in market model residuals, as the model in equation (9.33) is referred to in finance, has been reported by other researchers.

Fisher (1966) and Fama, Fisher, Jensen, and Roll (1969) reported average autocorrelation coefficients of −.12 and −.10,

TABLE 9.1 Initial RCR Estimates for the Market Model

Coefficient	Estimate of coefficient mean	Standard error of the estimate	Population standard deviation*	Test for mean zero, t	Test for randomness, χ^2
β_{oi}	0.0017	0.0005	*	3.4	36.9
β_{li}	1.2093	0.0492	0.3001	24.6	
1% critical value				2.7	99.3
Degrees of freedom				45	69

*Indicates negative estimate of population variance.

TABLE 9.2 Pooled Test for Serial Correlation in the Market Model

Pooled estimate $\hat{\rho}$	Standard error of the estimate	χ^2 -- statistic for simultaneous test of equality $H_0: \rho_1 = \cdots = \rho_N = 0$
-0.0858	0.0096	166.6
1% critical value		100.5
Degrees of freedom		70

respectively, when estimating the model using monthly data. Schwartz and Whitcomb (1977) support these findings for a sample of 20 firms. They also suggest a number of possible explanations for the presence of autocorrelated residuals including omission of explanatory variables, the impact of market markers and the Fisher effect. They find evidence supporting only the Fisher effect as a viable explanation.

In examining the effect of excluded variables they consider the market return variable lagged one, two and three time periods. In the repurchase study the possible exclusion of other important variables could be investigated.

The omission of certain significant variables from a regression can result in bias in the parameter estimates. Also, the omission of variables can result in serial correlation of the residuals in the models.

The model specified in equation (9.33) has two additional variables found to be significant, on average, for the 70 firms in the sample. The final model examined by Dielman et al. (1980) was

$$(R_{it} - R_{Ft}) = \beta_{0i} + \beta_{1i}(R_{mt} - R_{Ft}) + \beta_{2i}Z_{o;it} + \beta_{3i}S_{it} + \varepsilon_{it} \tag{9.34}$$

where the additional terms are

$Z_{o;it}$ = 1 if firm i announced a repurchase in time period t
= 0 otherwise

S_{it} = 0 in all months before the repurchase announcement
= $(R_{mt} - R_{Ft})$ in the month of announcement and all subsequent months.

The coefficient β_{2i} will represent any change in return in the month of announcement. The coefficient β_{3i} will represent any shift in the ith securities' beta coefficient (slope) due to the announcement. Note that the slope can now be written as

$$\text{slope} = \begin{cases} \beta_{1i} & \text{prior to the announcement} \\ \beta_{1i} + \beta_{3i} & \text{at and after the announcement} \end{cases}$$

The model in (9.34) was estimated by RCR with the results reported in Table 9.3.

The estimate of the coefficient mean for the c_i indicates an average increase of about $9\frac{1}{2}\%$ in rate of return during the announcement month. The population standard deviation does indicate, however, that there is significant variability in this increase.

As can be seen in Table 9.4, the inclusion of these two additional variables does little to change the results of the test for serial correlation. The value of the test statistic, 166.01, still exceeds the critical value, 100.5, at the 1% level of significance.

The analysis including the additional variables $Z_{o;it}$ and S_{it} still resulted in a negative estimate of variance for the intercept when all coefficients were assumed random.

A negative variance estimate in an RCR model can create many problems. Some of these problems are pointed out in Dielman (1980, pp. 40-51). Briefly,

1. Values of t-statistics for testing significance of parameters can be inflated.
2. Simultaneous tests of significance will be affected.
3. Tests for randomness will be affected.

It is suggested, therefore, that before any inferential conclusions are drawn the model should be reestimated with the variance constrained to equal zero for all coefficients that had

TABLE 9.3 RCR Estimates for the Announcement Effect Model

Coefficient	Estimate of coefficient mean	Standard error of the estimate	Population standard deviation*	Test for mean zero, t	Test for randomness, χ^2
β_0	0.0028	0.0007	*	4.0	
β_{1i}	1.1681	0.0482	0.3574	24.3	
β_{2i}	−0.1868	0.0421	0.2450	−4.4	138.8
β_{3i}	0.0951	0.0181	0.1292	5.2	204.8
1% critical value				2.6	99.3
Degrees of freedom				69	69

*The variance of β_{o1} has been constrained to equal zero.

TABLE 9.4 Pooled Test for Serial Correlation in the Announcement Effect Model

Pooled estimate $\hat{\rho}$	Standard error of the estimate	χ^2 -- statistic for simultaneous test of equality $H_0:\rho_1 = \cdots = \rho_N = 0$
−0.0834	0.0094	166.6
1% critical value		100.5
Degrees of freedom		70

initial negative variance estimates, that is, use the techniques from Chapter 5 for mixed RCR models. This has been done in both Table 9.3 and Table 9.5.

Mixed RCR estimates for the announcement model were obtained after differencing to remove serial correlation. These results are presented in Table 9.5. To perform the differencing the OLS residual estimates of the ρ_i were used for each firm. Thus one observation was lost for each firm. Due to the length of the time series it was felt this loss should be inconsequential.

After examining Table 9.5 it can be seen that none of the basic conclusions change concerning significance of variables in the announcement effect model. Thus serial correlation, although found to be significant using the pooled test, is not severe enough to seriously affect the RCR estimates of coefficient means.

The test for serial correlation can be applied to pooled models including Zellner's SUR and Swamy's RCR model. It is a simple test to apply, in that, computationally, it presents few problems. Also, the test could be extended easily to higher order schemes of autocorrelation or even to what Guilkey and Schmidt (1973) term a process with "vector autoregressive errors." Guilkey (1974) presented two possible tests for a first-order vector autoregressive process and compared them in a Monte Carlo experiment. Neither performed well for sample sizes less than 50; for larger samples, however, they were fairly reliable. Since these

TABLE 9.5 RCR Estimates for the Announcement Effect Model After Generalized Differencing

Coefficient	Estimate of coefficient mean	Standard error of the estimate	Population standard deviation*	Test for mean zero, t
β_0	0.0027	0.0006	*	4.3
β_{1i}	1.1657	0.0481	0.3599	24.2
β_{2i}	−0.1791	0.0412	0.2405	−4.3
β_{3i}	0.0970	0.0180	0.1289	5.4
1% critical value				2.6
Degrees of freedom				69

*The variance of β_{0i} has been constrained to equal zero.

tests can be used in certain of the same instances as the "pooled" test presented here, a Monte Carlo comparison might prove interesting. The pooled test is asymptotically justified (as are Guilkey's tests) but its small sample behavior is unknown.

Examination of RCR estimation results before and after differencing shows a few minor changes but nothing large enough to affect the significance of any test statistics calculated. The nondifferenced inferences reported in Dielman et al. (1980), still hold. This might suggest that RCR methods are fairly robust to the presence of a small degree of serial correlation in the disturbances as, for example, in the market model. The robustness of RCR estimates to problems such as serial correlation could be investigated by Monte Carlo simulation.

9.3 REGRESSION DIAGNOSTICS AND ROBUST ESTIMATION

When using separate equation regressions, the presence of a single unusual observation (outlier) can have a significant effect on the OLS estimates. Procedures are available in the separate equation case to help in detecting which observations might be classified as outliers or which might be influential in determining the fit of the regression surface. Many of these procedures are discussed in Belsley, Kuh, and Welsch (1980) and Cook and Weisberg (1982).

Outliers could also be expected to be influential in determining the fit of pooled models. Thus, to develop measures of influence for pooled models would appear to be of interest to researchers. Certain of the measures already developed could be extended directly to models such as the CP model of Chapter 2. For other models, such as the RCR model, for example, modifications might be necessary since the mean of the coefficients pertaining to a number of individuals is being estimated. A single outlier for an individual might affect the OLS etimates of that individual's coefficients but its effect could be much less a factor in determining the estimate of the mean for all individuals.

The population mean regression coeficient, $\bar{\beta}$, estimated using Swamy's RCR procedure is a weighted average of the OLS estimates of the individual coefficients:

$$\hat{\bar{\beta}} = \sum_{i-1}^{N} W_i\hat{\beta}_i \tag{9.35}$$

where W_i is defined in equation (5.9) and the $\hat{\beta}_i$ are the separate equation OLS estimates of the coefficients. The estimates, $\hat{\beta}_i$, may be affected by outliers within the ith time series of observations as mentioned previously. Further, the average, $\hat{\bar{\beta}}$, may be affected by values of the $\hat{\beta}_i$ that differ greatly from most of the OLS estimates. Thus outlier detection procedures might be better applied to the $\hat{\beta}_i$, treated as a sample of N observations. The weighting of the $\hat{\beta}_i$ would have to be taken into account, however. Since each coefficient is weighted in inverse relation to its variance, the weighting may result in some reduction is the effect of extreme $\hat{\beta}_i$ values in determining $\hat{\bar{\beta}}$.

In separate equation regression there are two primary ways of treating the possible problems caused by the presence of outliers. To reduce the effect of outliers on coefficient estimates one might: delete the outliers, or, use a robust estimation method which is not as strongly affected as OLS. The second alternative includes such methods as L_p-norm estimation, M-estimation, R-estimation and bounded influence regression. See Huber (1972, 1973, 1981), Koenker (1982), Dielman (1984) and Dielman and Pfaffenberger (1982) for surveys of various robust estimation techniques and references. These robust methods could possible be incorporated into pooled models.

9.4 MODELS WITH TIME AND CROSS-SECTIONALLY VARYING COEFFICIENTS

In Chapter 6 models were discussed which allowed coefficients to vary over time and individuals. These models have seen little use in practice due in part to their complexity and to the lack of available computer programs to implement the estimation and hypothesis testing procedures. It would seem that such general models could prove useful in practice as suggested by preliminary empirical analyses in Swamy and Mehta (1977a).

To date, little is known of how models with both time and cross-sectional variation in the coefficients will perform in small samples. In large samples, however, asymptotic results suggest that the models should perform well.

A very general model of coefficient variation over time has been developed by Swamy and Tinsley (1980). As yet, this model has not been extended to cases where time series observations are available for a number of individuals. This extension could be made however, as noted by Dielman (1983). The same problems with computational complexity would be present but the complexity of the model does not preclude identification and consistent estimation of the model parameters. See also Havenner and Swamy (1981).

9.5 OTHER RESEARCH OPPORTUNITIES

One of the primary hindrances to the analysis of pooled cross-sectional and time series data for many researchers is the lack of available software for implementing certain of the techniques. This was noted in particular for the time and cross-sectionally varying coefficient models discussed in Chapter 6. Although programs for most other procedures can be found, there is no software package known to the author which incorporates a full range of the procedures. The most comprehensive package found was the TSCS package which is a subset of TSP, an econometric package. It would be useful if statistical packages such as SAS or SPSS could incorporate such routines. Alternatively, perhaps IMSL (1984) could incorporate subroutines which would make the writing of FORTRAN programs to implement the procedures easier.

A variety of other articles appearing in the recent literature is reflective of the varied research opportunities in analysis of pooled cross-sectional and time series data. For examples, see Deaton (1985), Pakes and Griliches (1984), Ware (1985), Heckman and Robb (1985), Singer (1981), Singer and Cohen (1980), Singer and Spilerman (1976), Trivedi (1980), Maddala (1978), Griliches, Hall, and Hausman (1978), Glejser (1978), and Beggs (1986).

In addition, the interested reader is referred to the article by Hsiao (1985) where problems of measurement error, development of discrete choice models, and the use of dynamic models are discussed. Although Hsiao presents these topics in an error components model framework many of the research avenues he suggests are also pertinent to other models for pooled cross-sectional and time series data.

Appendix A

PROOF THAT THE CP ESTIMATOR IS UNBIASED WHEN COEFFICIENTS ARE RANDOM

Consider the RCR model of Chapter 5 written as in equation (5.1)

$$Y_i = X_i\beta_i + \varepsilon_i \tag{A.1}$$

where

$$\beta_i = \bar{\beta} + v_i \tag{A.2}$$

Equation (A.1) can be written as

$$Y_i = X_i(\bar{\beta} + v_i) + \varepsilon_i \tag{A.3}$$

$$= X_i\bar{\beta} + X_iv_i + \varepsilon_i \tag{A.4}$$

$$= X_i\bar{\beta} + \varepsilon_i \tag{A.5}$$

where

$$e_i = X_iv_i + \varepsilon_i \tag{A.6}$$

Under the model assumptions it is true that $E(e_i) = 0$.

Suppose the CP estimator $\hat{\bar{\beta}}_{CP1}$ from equation (2.5) is used to estimate $\bar{\beta}$. The estimator can be rewritten as

$$\hat{\bar{\beta}}_{CP1} = (Z'Z)^{-1}Z'Y \tag{A.7}$$

$$= \left(\sum_{i=1}^{N} X_i'X_i \right)^{-1} \sum_{i=1}^{N} X_i'Y_i \tag{A.8}$$

$$= \left(\sum_{i=1}^{N} X_i'X_i \right)^{-1} \sum_{i=1}^{N} X_i'(X_i\bar{\beta} + e_i) \tag{A.9}$$

$$= \left(\sum_{i=1}^{N} X_i'X_i \right)^{-1} \sum_{i=1}^{N} X_i'X_i\bar{\beta} + \left(\sum_{i=1}^{N} X_i'X_i \right)^{-1} \sum_{i=1}^{N} X_i'e_i \tag{A.10}$$

$$= \bar{\beta} + \left(\sum_{i=1}^{N} X_i'X_i \right)^{-1} \sum_{i=1}^{N} X_i'e_i \tag{A.11}$$

Since $E(e_i) = 0$, we have

$$E(\hat{\bar{\beta}}_{CP1}) = \bar{\beta}$$

Thus, if the coefficients are random, the CP estimator will be an unbiased estimator of $\bar{\beta}$, the mean of the random coefficients.

Appendix B

DERIVATION OF VARIABLE SELECTION RESULTS FOR THE GENERAL LINEAR REGRESSION MODEL

"True" Model:

$$Y = X\beta + \varepsilon \tag{B.1}$$

$$= X_P\beta_P + X_R\beta_R + \varepsilon \tag{B.2}$$

where

$$X = [X_P \; X_R] \tag{B.3}$$

$$\beta = \begin{bmatrix} \beta_P \\ \beta_R \end{bmatrix} \tag{B.4}$$

$$E(\varepsilon) = 0 \tag{B.5}$$

and

$$E(\varepsilon\varepsilon') = \Omega \tag{B.6}$$

Misspecified Model:

$$Y = X_P\beta_P + \varepsilon \tag{B.7}$$

where

$$E(\varepsilon) = 0 \tag{B.8}$$

and

$$E(\varepsilon\varepsilon') = \Omega \tag{B.9}$$

Estimation

True Model:

$$\tilde{\beta} = \begin{bmatrix} \tilde{\beta}_P \\ \tilde{\beta}_R \end{bmatrix} = (X'\Omega^{-1}X)^{-1}X'\Omega^{-1}Y \tag{B.10}$$

Misspecified Model:

$$\tilde{\beta}_P^* = (X_P'\Omega^{-1}X_P)^{-1}X_P'\Omega^{-1}Y \tag{B.11}$$

Bias in $\tilde{\beta}_P^*$

$$\tilde{\beta}_P^* = (X_P'\Omega^{-1}X_P)^{-1}X_P'\Omega^{-1}Y \tag{B.12}$$

$$= (X_P'\Omega^{-1}X_P)^{-1}X_P'\Omega^{-1}(X_P\beta_P + X_R\beta_R + \varepsilon) \tag{B.13}$$

$$= \beta_P + (X_P'\Omega^{-1}X_P)^{-1}X_P'\Omega^{-1}X_R\beta_R$$
$$+ (X_P'\Omega^{-1}X_P)^{-1}X_P'\Omega^{-1}\varepsilon \tag{B.14}$$

$$E(\tilde{\beta}_P^*) = \beta_P + (X_P'\Omega^{-1}X_P)^{-1}X_P'\Omega^{-1}X_R\beta_R \tag{B.15}$$

$$= \beta_P + B \tag{B.16}$$

Thus, unless $\beta_R = 0$, the estimator $\tilde{\beta}_P^*$ will have bias equal to the term B shown in equations (B.15) and (B.16).

Variance of $\tilde{\beta}_P^*$ and $\tilde{\beta}_P$

$$\text{Var}(\tilde{\beta}_P^*) = (X_P'\Omega^{-1}X_P)^{-1} \tag{B.17}$$

$$\text{Var}(\tilde{\beta}_P) = \text{appropriate partition of } (X'\Omega^{-1}X)^{-1} \tag{B.18}$$

where

$$(X'\Omega^{-1}X)^{-1} = \begin{bmatrix} X_P'\Omega^{-1}X_P & X_P'\Omega^{-1}X_R \\ X_R'\Omega^{-1}X_P & X_R'\Omega^{-1}X_R \end{bmatrix}^{-1} \tag{B.19}$$

By an identity for inversion of a partitioned matrix given in Theil (1971, p.18):

$$(X'\Omega^{-1}X)^{-1} = \begin{bmatrix} C & D \\ E & F \end{bmatrix} \tag{B.20}$$

where

$$\begin{aligned} C = {} & (X_P'\Omega^{-1}X_P)^{-1} + (X_P'\Omega^{-1}X_P)^{-1}X_P'\Omega^{-1}X_R \\ & \times [X_R'\Omega^{-1}X_R - X_R'\Omega^{-1}X_P(X_P'\Omega^{-1}X_P)^{-1}X_P'\Omega^{-1}X_R]^{-1} \\ & \times X_P'\Omega^{-1}X_R(X_P'\Omega^{-1}X_P)^{-1} \end{aligned} \tag{B.21}$$

Thus,

$$\text{Var}\ (\tilde{\beta}_P) = (X_P'\Omega^{-1}X_P)^{-1} + W \tag{B.22}$$

where W = the second term in equation (B.21) and is a psd matrix.

Therefore Var $(\tilde{\beta}p)$ − Var $(\tilde{\beta}p^*)$ = is psd, that is, the estimates of the components of βp given by $\tilde{\beta}p$ are generally more variable than those given by $\tilde{\beta}p^*$.

Mean Square Error (MSE) of $\tilde{\beta}p^*$

$$\text{MSE}\ (\tilde{\beta}_P^*) = \text{Var}\ (\tilde{\beta}_P^*) + B^2 \tag{B.23}$$

where B is the bias given in equation (B.15).

Writing

$$A = (X_P'\Omega^{-1}X_P)^{-1}X_P'\Omega^{-1}X_R \tag{B.24}$$

then

$$\text{MSE}\ (\tilde{\beta}_P^*) = (X_P'\Omega^{-1}X_P)^{-1}\ A\beta_R\beta_R'A' \tag{B.25}$$

It follows that

$$\text{Var}\ (\tilde{\beta}_P) - \text{MSE}\ (\tilde{\beta}_P^*) = A[X_R'\Omega^{-1}X_R - (X_R'\Omega^{-1}X_P)A]^{-1} \times A' - A\beta_R\beta_R'A' \tag{B.26}$$

Also, note that

$$\text{Var}\ (\tilde{\beta}_R) - \beta_R\beta_R' = [X_R'\Omega^{-1}X_R - X_R'\Omega^{-1}X_P(X_P'\Omega^{-1}X_P)^{-1} \times X_P'\Omega^{-1}X_R]^{-1} - \beta_R\beta_R' \tag{B.27}$$

$$= [X_R'\Omega^{-1}X_R - (X_R'\Omega^{-1}X_P)A]^{-1} - \beta_R\beta_R' \tag{B.28}$$

So

$$\text{Var}\ (\tilde{\beta}_P) - \text{MSE}\ (\tilde{\beta}_P^*) = A\ [\text{Var}\ (\tilde{\beta}_R) - \beta_R\beta_R')A' \tag{B.29}$$

Therefore if Var $(\tilde{\beta}_R) - \beta_R\beta_R'$ is psd, then so is Var $(\tilde{\beta}_P)$ − MSE $(\tilde{\beta}_P^*)$.

References

Akkina, K. R. (1974), Application of random coefficient regression models to the aggregation problem, Econometrica, 42, 369-375.

Albon, R. P. and Valentine, T. J. (1977), The sectoral demand for bank loans in Australia, Econom. Record, 53, 167-181.

Amemiya, T. (1985), Advanced Econometrics, Cambridge: Harvard University Press.

___ (1981), Qualitative response models: A survey, J. Econom. Literature, 19, 1483-1536.

___ (1978), A note on a random coefficient model, Internat. Econom. Rev., 19, 793-796.

___ (1971), The estimation of the variances in a variance-components model, Internat. Econom. Rev., 12, 1-13.

Amemiya, T. and MaCurdy, T. E. (1986), Instrumental-variable estimation of an error-components model, Econometrica, 54, 869-880.

Anderson, T. W. and Hsiao, C. (1982), Formulation and estimation of dynamic models using panel data, J. Econometrics, 18, 47-82.

___ (1981), Estimation of dynamic models with error components, J. Amer. Statist. Assoc., 76, 598-606.

___ (1971), Estimation of covariance matrices with linear structure and moving average processes of finite order, Technical Report No. 6, Department of Statistics, Stanford University.

Archibald R. and Gillingham, R. (1980), An analysis of the short-run consumer demand for gasoline using household survey data, Rev. Econom. Statist., 62, 622-628.

Arora, S. S. (1973), Error components regression models and their applications, Ann. Econom. Social Measurement, 2, 451-561.

Ashenfelter, O. (1978), Estimating the effect of training programs on earnings, Rev. Econom. Statist., 60, 47-57.

Atkinson, M. and Mairesse, J. (1978), Length of life of equipment in French manufacturing industries, Ann. I.N.S.E.E., 30-31, 23-48.

Avery, R. (1977), Error components and seemingly unrelated regressions, Econometrica, 45, 199-209.

Avery, R. B., Hansen, L. P., and Hotz, V. J. (1983), Multi-period probit models and orthogonality condition estimation, Internat. Econom. Rev., 24, 21-35.

Baksalary, J. K. and Kala, R. (1979), On the prediction problem in the seemingly unrelated regression equations model, Math. Operationsforsch. Statist. Ser. Statist., 10, 203-208.

Balestra, P. and Nerlove, M. (1966), Pooling cross section and time series data in the estimation of a dynamic model: The demand for natural gas, Econometrica, 34, 585-612.

Baltagi, B. H. (1986), Pooling under misspecification: Some Monte Carlo evidence on the Kmenta and the error components techniques, Econometric Theory, 2, 429-440.

___ (1984), A Monte Carlo study of pooling time-series and cross-section data in the simultaneous equations model. Internat. Econom. Rev., 25, 603-624.

___ (1981a), Pooling: An experimental study of alternative testing and estimation procedures in a two-way error component model, J. Econometrics, 17, 21-49.

___ (1981b), Simultaneous equations with error components, J. Econometrics, 17, 89-200.

___ (1980), On seemingly unrelated regressions with error components, Econometrica, 48, 1547-1551.

Baltagi, B. H. and Griffin, J. M. (1984), Short and long run effects in pooled models, Internat. Econom. Rev., 25, 631-645.

___ (1983), Gasoline demand in the OECD: An application of pooling and testing procedures, European Econom. Rev., 22, 117-137.

Barth, J., Kraft, A., and Kraft, J. (1979), A temporal cross-section approach to the price equation, J. Econometrics, 11, 335-351.

Bass, F. M. and Wittink, D. R. (1978), Pooling issues and methods in regression analysis: Some further reflections, J. Marketing Res., 15, 277-279.

___ (1975), Pooling issues and methods in regression analysis with examples in marketing research, J. Marketing Res., 12, 414-425.

Beach, C. M. and MacKinnon, J. G. (1979), Maximum likelihood estimation of singular equation systems with autoregressive disturbances. Internat. Econom. Rev., 20, 459-464.

Becker, W. E. and Morey, M. J. (1980), Pooled cross-section, time-series evaluation: Source, result, and correction of serially correlated errors, Amer. Econom. Rev., 70, 35-40.

Beckwith, N. (1972), Multivariate analysis of sales response of competing brands to advertising, J. Marketing Res., 9, 168-176.

Beggs, J. (1986), Time series analyses in pooled cross-sections, Econometric Theory, 2, 331-349.

Beierlein, J. G., Dunn, J. W., and McConnon, J. C., Jr. (1981), The demand for electricity and natural gas in the northeastern United States, Rev. Econom. Statist., 63, 403-408.

Belsley D. A., Kuh, E., and Welsch, R. E. (1980), Regression Diagnostics: Identifying Influential Data and Sources of Collinearity, New York: Wiley.

Benus, J., Kmenta, J., and Shapiro, H. (1976), The dynamics of household budget allocation to food expenditures, Rev. Econom. Statist., 58, 129-138.

Berk, R. A., Hoffman, D. M., Maki, J. E., Rauma, D., and Wong, H. (1979), Estimation procedures for pooled cross-sectional and time series data, Evaluation Quart., 3, 385-410.

Berndt, E. and Savin, N. (1977), Conflict among criteria for testing hypotheses in the multivariate regression model, Econometrica, 45, 1263-1278.

Berry, M. and Trennepohl, G. L. (1981), A comparative analysis of security performance methodology, Bureau of Business and Economic Research Working Paper, No. FN 81-10, Arizona State Univeristy.

Berzeg, K. (1982), Demand for motor gasoline: A generalized error components model, Southern Econom. J., 49, 462-471.

___ (1979), The error compnents model: Conditions for the existence of the maximum likelihood estimates, J. Econometrics, 10, 99-102.

Bhargava, A. and Sargan, J. D. (1983), Estimating dynamic random effects models from panel data covering short time periods, Econometrica, 51, 1635-1659.

Binder, J. J. (1983), Measuring the Effects of Regulation with Stock Price Data: A New Methodology, Unpublished Ph.D. Dissertation, University of Chicago.

Binkley, J. (1982), The effect of variable correlation on the efficiency of seemingly unrelated regression in a two-equation model, J. Amer. Statist. Assoc., 77, 890-895.

Biorn, E. (1981), Estimating economic relations from incomplete cross-section/time-series data, J. Econometrics, 16, 221-236.

Blair, R. and Kraft, J. (1974), Estimation of elasticity of substitution in American manufacturing industry from pooled cross-section and time-series observations, Rev. Econom. Statist., 56, 343-347.

Boness, J. and Frankfurter, G. (1977), Evidence of non-homogeneity of capital costs within risk classes, J. Finance, 32, 775-787.

Boot, J. and Frankfurter, G. (1972), The dynamics of corporate debt management, decision rules, and some empirical evidence, J. Fin. Quant. Analysis, 7, 1957-1965.

Boot, J. C. G. and deWit, G. M. (1960), Investment demand: An empirical contribution to the aggregation problem, Internat. Econom. Rev., 1, 3-30.

Brehm, C. T. and Saving, T. R. (1964), The demand for general assistance payments, Amer. Econom. Rev., 54, 1002-1018.

Breusch, T. S. and Pagan, A. R. (1980), The Lagrange multiplier test and its application to model specification in econometrics, Rev. Econom. Stud., 47, 239-253.

Brobst, R. and Gates, R. (1977), Comments on pooling issues and methods in regression analysis, J. Marketing Res., 14, 598-600.

Brown, K. and Kadiyala, K. (1985), The estimation of missing observation in related time series data: Further results, Commun. Statist.—Simula. Computa., 14, 973-981.

___ (1983), construction of economic index numbers with an incomplete set of data, Rev. Econom. Statist., 65, 520-524.

Buse, A. (1979), Goodness-of-fit in the seemingly unrelated regressions model: A generalization, J. Econometrics, 10, 109-113.

Carlson, R. (1978), Seemingly unrelated regression and the demand for automobiles of different sizes, 1965-1975: A disaggregate approach, J. Business, 51, 243-262.

Carlson R. L. and Umble, M. M. (1980), Statistical demand functions for automobiles and their use for forecasting in an energy crisis, J. Business, 53, 193-204.

Carter, R. L. and Yang, M. C. K. (1986), Large sample inference in random coefficient regression models Comm. Statist. A-Theory Methods, 15, 2507-2525.

Chamberlain, G. (1985), Heterogeneity, omitted variable bias, and duration dependence, in Longitudinal Analysis of Labor Market Data, (Eds. J. J. Heckman and B. Singer) Cambridge University Press, 3-38.

___ (1984), Panel data, in Handbook of Econometrics, Vol. 2, (Eds. Z. Griliches and M. Intriligator) Amsterdam: North-Holland, 1247-1318.

___ (1982), Multivariate regression models for panel data, J. Econometrics, 18, 5-46.

___ (1980), Analysis of covariance with qualitative data, Rev. Econom. Stud., 47, 225-238.

___ (1978), Omitted variable bias in panel data: Estimating the returns to schooling, Ann. I.N.S.E.E., 30-31, 49-82.

Chang, Y. (1978), A study of industry location from pooled time-series and cross-section data: The case of cotton textile mills, Quart. Rev. Econom. Business, 19, 75-88.

Chang, H. and Lee, C. (1977), Using pooled time-series and cross-section data to test the firm and time effects in financial analyses, J. Fin. Quant. Analysis, 12, 457-471.

Chetty, V. K. (1968), Pooling of time series and cross section data, Econometrica, 36, 279-289.

Condie, J. (1977), A very high level language for econometric research, Econometrica, 45, 238-239.

Conniffe, D. (1985), Estimating regression equations with common explanatory variables but unequal numbers of observations, J. Econometrics, 27, 179-196.

___ (1982a), Testing the assumptions of seemingly unrelated regressions, Rev. Econom. Statist., 64, 172-174.

___ (1982b), Covariance analysis and seemingly unrelated regressions, Amer. Statist., 36, 169-171.

Cook, R. D. and Weisberg, S. (1982), Residuals and Influence in Regression, New York: Chapman and Hall.

Daganzo, C. F. and Sheffi, Y. (1982), Multinomial probit with time-series data: Unifying state dependence and serial correlation models, Environment and Planning A, 14, 1377-1388.

Da Silva, J. G. C. (1975), The analysis of cross-sectional time series data, Institute of Statistics Mimeograph Series No. 1011, Raleigh, North Carolina, North Carolina State University, unpublished Ph.D. dissertation.

Davies, R. B. and Crouchley, R. (1984), Calibrating longitudinal models of residential mobility and migration, J. Regional Sci. Urban Econom., 14, 231-247.

Deaton, A. (1985), Panel data from time series of cross-sections, J. Econometrics, 30, 109-126.

Desai, M. (1974), Pooling as a specification error—a note, Econometrica, 42, 389-391.

Dhrymes, P. J. (1971), Equivalence of iterative Aitken and maximum likelihood estimators for a system of regression equations, Australian Econom. Papers, 10, 20-24.

Dielman, T. E. (1984), Least absolute value estimation in linear regression: An annotated bibliography, Comm. Statist.—Statistical Rev., 13, 513-541.

___ (1983), Pooled cross-sectional and time series data: A survey of current statistical methodology, Amer. Statistician, 37, 111-122.

___ (1980), Pooled Data for Financial Markets (Research for Business Decision Series), Ann Arbor: UMI Research Press.

Dielman, T. E. and Nantell, T. J. (1982), Tender offer mergers and stockholders' wealth: A random coefficient regression approach, Amer. Statist. Assoc. Business Econom. Statist. Proc., 334-339.

Dielman, T. E. and Oppenheimer, H. R. (1984), An examination of investor behavior during periods of large dividend changes, J. Fin. Quant. Analysis, 19, 197-216.

Dielman, T. E. and Pfaffenberger, R. C. (1982), LAV (least absolute value) estimation in linear regression: A review, in TIMS Studies in the Management Sciences, (Ed. S. Zanakis and J. Rustagi), Amsterdam, North-Holland, 19, 31-52.

Dodd, P. (1980), Merger proposals, management discretion, and stockholder wealth, J. Finan. Econom., 8, 105-138.

Dodd, P. and Ruback, R. (1977), Tender offers and stockholder returns: An empirical analysis, J. Finan. Econom., 5, 351-374.

Don, F. J. H. and Magnus, J. R. (1980), On the unbiasedness of iterated GLS estimators, Comm. Statist. A-Theory Methods, A9, 519-527.

Doran, H. E. and Griffith, W. E. (1983), On the relative efficiency of estimators which include the initial observations in the estimation of seemingly unrelated regressions with first-order autoregressive disturbances, J. Econometrics, 23, 165-191.

Drummond, D. J. and Gallant, A. R. (1977), TSCSREG: A SAS procedure for the analysis of time series cross-section data, Institute of Statistics Mimeograph Series No. 1107, Raleigh, North Carolina, North Carolina State University.

Duncan, G. M. (1983), Estimation and inference for heteroskedastic systems of equations, Inernat. Econom. Rev., 24, 559-566.

Durbin, J. (1953), A note on regression when there is extraneous information about one of the coefficients, J. Amer. Statist. Assoc., 48, 799-808.

Dwivedi, T. D. and Srivastava, V. K. (1978), Optimality of least squares in the seemingly unrelated regression equation model, J. Econometrics, 7, 391-395.

Eisner, R. (1978), Cross section and time series estimates of investment functions, Ann. I.N.S.E.E., 30-31, 99-129.

Fama, E. F., Fisher, L., Jensen, M. C., and Roll, R. (1969), The adjustment of stock prices to new information, Internat. Econom. Rev., 10, 1-21.

Fanfani, R. (1975), Pooling time series and cross-section data: A review, European Rev. Agricultural Econom., 2, 65-85.

Faurot, D. J. and Fon, V. (1978), A computer program for seemingly unrelated nonlinear regressions, Econometrica, 46, 479.

Fearn, T. (1975), A Bayesian approach to growth curves, Biometrica, 62, 89-100.

Feige, E. L. (1974), Temporal cross-section specifications of the demand for demand deposits, J. Finance, 29, 923-940.

Feige E. and Swamy, P.A.V.B. (1974), A random coefficient model of the demand for liquid assets, J. Money Credit Banking, 6, 241-252.

Ferguson, I. and Leech, J. (1978), Generalized least squares estimation of yield functions, Forest Sci., 24, 27-42.

Fisher, L. (1966), Some new stock market indexes, J. Business, 39, 191-225.

Fisk, P. R. (1967), Models of the second kind in regression analysis, J. Roy. Statist. Soc., Ser. B, 29, 266-281.

Flinn, C. and Heckman, J. J. (1983a), The likelihood function for the multistate-multiepisode model in 'Models for the analysis of labor force dynamics,' in Advances in Econometrics, Vol. 2, (Eds. R. Bassmann and G. Rhodes) Greenwich, Conn.: JAI Press, Inc., 225-231.

___ (1983b), Erratum and addendum to volume 1: Models for the analysis of labor force dynamics, in Advances in Econometrics, Vol. 2, (Eds. R. Bassmann and G. Rhodes) Greenwich, Conn.: JAI Press, Inc., 219-223.

___ (1982a), Models for the analysis of labor force dynamics, in Advances in Econometrics, Vol. 1 (Eds. R. Bassmann and G. Rhodes) Greenwich, Conn.: JAI Press Inc., 35-95.

___ (1982b), New methods for analyzing structural models of labor force dynamics, J. Econometrics, 18, 115-168.

Fomby, T. B., Hill, R. C., and Johnson, S. R. (1984), Advanced Econometric Methods, New York: Springer-Verlag.

Frieden, A. (1973), A program for the estimation of dynamic economic relations from a time series of cross sections, Ann. Econom. Social Measurement, 2, 89-91.

Fuller, W. and Battese, G. E. (1974), Estimation of linear models with crossed-error structure, J. Econometrics, 2, 67-78.

___ (1973), Transformations for estimation of linear models with nested-error structure, J. Amer. Statist. Assoc., 68, 626-632.

Gahlon, J. and Stover, R. (1979), Diversification, financial leverage, and conglomerate systematic risk, J. Finan. Quant. Analysis, 14, 999-1014.

Gallant, R. (1975), Seemingly unrelated nonlinear regressions, J. Econometrics, 3, 35-50.

Gendreau, B. C. and Humphrey, D. B. (1980), Feedback effects in the market regulation of bank leverage: A time-series and cross-section analysis, Rev. Econom. Statist., 62, 276-280.

Ghalai, M. A. (1977), Pooling as a specification error: A comment, Econometrica, 755-757.

Gibbons, M. R. (1982), Multivariate tests of financial models: A new approach, J. Fin. Econom., 10, 3-27.

Glejser, H. (1978), Truncated distributed lags in small sample panel models, Ann. I.N.S.E.E., 30-31, 131-135.

Goodnight, J. and Wallace, T. D. (1972), Operational techniques and tables for making weak MSE tests for restrictions in regressions, Econometrica, 40, 699-709.

Granger, C., Engle, R., Ramanathan, R., and Andersen, A. (1979), Residential load curves and time-of-day pricing: An econometric analysis, J. Econometrics, 9, 13-32.

Greene, D. L. (1980), Regional demand for gasoline: Comment, J. Regional Sci., 20, 103-109.

Green, H. A. J. (1964), Aggregation in Economic Analysis: An Introductory Survey, Princeton: Princeton University Press.

Griffin, J. (1977), Inter-fuel substitution possibilities: A translog application of intercountry data, Internat. Econom. Rev., 18, 755-770.

Griffiths, W. E. (1974), Combining time series cross-section data: Alternative models and estimators, Paper presented at the Fourth Conference of Economists, Canberra.

Griliches, Z. (1977), Estimating the returns to schooling: Some economic problems, Econometrica, 45, 1-22.

Griliches, Z. and Hausman, J. (1986), Errors in variables in panel data, J. Econometrics, 31, 93-118.

Griliches, J. Hall, B. H., and Hausman, J. A. (1978), Missing data and self-selection in large panels, Ann. I.N.S.E.E., 30-31, 137-176.

Guilkey, D. (1974), Alternative tests for a first-order vector autoregressive error specification, J. Econometrics, 2, 95-104.

Guilkey, D. K. and Schmidt, P. (1973), Estimation of seemingly unrelated regressions with vector autoregressive errors, J. Amer. Statist. Assoc., 68, 642-647.

Haavelmo, T. (1947), Family expenditures and the marginal propensity to consume, Econometrica, 15, 335-341.

Hall, H. D. (1982), The relative efficiency of time and frequency domain estimators in SUR systems, J. Statist. Comput. Simulation, 16, 81-96.

Hall, B. (1978), A general framework for the time series—cross section estimation, Ann. I.N.S.E.E., 30-31, 177-202.

Hannan, M. and Young, A. (1977), Estimation in panels models: Results on pooling cross-sections and time series, in Sociological Methodology, (Ed. D. R. Heise), San Francisco: Jossey-Bass Inc., 52-83.

Halpern, P. (1973), Empirical estimates of the amount and distribution of gains to companies in mergers, J. Business, 46, 554-575.

Harvey, A. (1982), A test of misspecification for systems of equations, Discussion Paper No. A31, London School of Economics Econometrics Programme, London, England.

___ (1978), The estimation of time-varying parametrs from panel data, Ann. I.N.S.E.E., 30-31, 203-226.

Harvey, A. C. and Phillips, G.D.A. (1982), Testing for contemporaneous correlation of disturbances in systems of regression equations, Bulletin of Econom. Res., 34, 79-91.

Harville, D. A. (1977), Maximum likelihood approaches to variance component estimation and to related problems, J. Amer. Statist. Assoc., 72, 320-338.

___ (1976), Extension of the Gauss-Markov Theorem to include the estimation of random effects, Ann Statist., 4, 383-395.

Hausman, J. A. (1978), Specification tests in econometrics, Econometrica, 46, 1251-1271.

Hausman, J. A. and Taylor, W. E. (1981), Panel data and unobservable individual effects, Econometrica, 49, 1377-1398.

Hausman, J. A. and Wise, D. A. (1979), Attrition bias in experimental and panel data: The Gary income maintenance experiment, Econometrica, 47, 455-473.

Havenner, A. and Herman, R. (1977), Pooled time-series cross-section estimation, Econometrica, 45, 1535-1536.

Havenner, A. and Swamy, P. A. V. B. (1981), A random coefficient approach to seasonal adjustment of economic time series, J. Econometrics, 15, 177-210.

Heckman, J. J. (1981a), Statistical models for discrete panel data, in Structural Analysis of Discrete Data with Econometric Applications, (Eds. C. F. Manski and D. McFadden), Cambridge, Mass.: The MIT Press, 114-178.

___ (1981b), The incidental parameters problem and the problem of initial conditions in estimating a discrete time—discrete data stochastic process, in Structural Analysis of Discrete Data with Econometric Applications, (Eds. C. F. Manski and D. McFadden), Cambridge, Mass.: The MIT Press, 179-195.

___ (1981c), Heterogeneity and state dependence, in Studies in Labor Markets, (Ed. S. Rosen) Chicago: The University of Chicago Press, 91-139.

___ (1978), Simple statistical models for discrete panel data developed and applied to test the hypothesis of true state dependence against the hypothesis of spurious state dependence, Ann. I.N.S.E.E., 30-31, 227-269.

Heckman, J. J. and Borjas, G. J. (1980), Does unemployment cause future unemployment? Definitions, questions and answers from a continuous time model of heterogeneity and state dependence, Economica, 47, 247-283.

Heckman, J. J. and MaCurdy, T. E. (1980), A life cycle model of female labour supply, Rev. Econom. Stud., 47, 47-74.

Heckman, J. J. and Robb. R. (1985), Alternative methods for evaluating the impact of interventions, in Longitudinal Analysis of Labor Market Data, (Eds. J. J. Heckman and B. Singer) Cambridge: Cambridge University Press, 156-245.

Heckman, J. J. and Singer, B. (1985), Social science duration analysis, in Longitudinal Analysis of Labor Market Data, (Eds. J. J. Heckman and B. Singer) Cambridge: Cambridge University Press, 39-110.

___ (1984a), Econometric duration analysis, J. Econometrics, 24, 63-132.

___ (1984b), A method of minimizing the impact of distributional assumptions in econometric models for duration data, Econometrica, 52, 271-320.

___ (1984c), The identifiability of the proportional hazard model, Rev. of Econom. Studies, 51, 231-241.

___ (1982), The identification problem in econometric models for duration data, in Advances in Econometrics, Proceedings of World Meetings of the Econometric Society, 1980 (Ed. W. Hildebrand) Cambridge: Cambridge University Press, 39-77.

Heckman, J. J. and Willis, R. J. (1977), A beta-logistic model for the analysis of sequential labor force participation by married women, J. Political Econom., 85, 27-58.

___ (1975), Estimation of a stochastic model of reproduction: An econometric approach, in Household Production and Consumtion, (Ed. N. E. Terleckyj), New York: National Bureau of Economic Research, 99-138 (with comments, pp. 139-145).

Henderson, C. (1971), Comment on the use of error components models in combining cross section with time series data, Econometrica, 39, 397-401.

Hendricks, W., Koenker, R., and Poirier, D. (1979), Stochastic parameter models for panel data: An application to the Connecticut peak load pricing experiment, Internat. Econom. Rev., 20, 707-724.

Henin, P. (1978), Individual and time effects in the dividend behavior of firms, Ann. I.N.S.E.E., 30-31, 271-296.

Henry, N. W., McDonald, J. F., and Stokes, H. H. (1976), The estimation of dynamic economic relations from a time series of cross sections: A programming modification, Ann. Econom. Social Measurement, 5, 153-155.

Hildreth, C. and Houck, J. P. (1968), Some estimates for a linear model with random coefficients, J. Amer. Statist. Assoc., 63, 584-595.

Hillier, G. H. and Satchell, S. E. (1986), Finite-sample properties of a two-stage single equation estimator in the SUR model, Econometric Theory, 2, 66-74.

Hirschey, M. (1981), The effect of advertising on industrial mobility, 1947-72, J. Business, 54, 329-339.

Hoch, I. (1962), Estimation of production function parameters combining time-series and cross-section data, Econometrica, 30, 34-53.

Hocking, R. R. (1976), The analysis and selection of variables in linear regression, Biometrics, 32, 1-50.

Houthakker, H. S., Verleger, P. K. Jr., and Sheehan, D. (1974), Dynamic demand for gasoline and residential electricity, Amer. J. Agricultural Econom., 56, 412-418.

Hsiao, C. (1986), Analysis of Panel Data, Cambridge: Cambridge Univeristy Press.

___ (1985), Benefits and limitations of panel data, Econometric Rev., 4, 121-174.

___ (1975), Some estimation methods for a random coefficient model, Econometrica, 43, 305-325.

___ (1974), Statistical inference for a model with both random cross-sectional and time effects, Internat. Econom. Rev., 15, 12-30.

___ (1972), The combined use of cross-section and time series data in econometric analysis, Umpublished Ph.D. Dissertation, Stanford University.

Huang, W. (1987), A pooled cross-section and time-series study of professional indirect immigration to the United States, Southern Econom. J., 54, 95-109.

Huber, P. J. (1981), Robust statistics, New York: John Wiley and Sons, Inc.

___ (1973), Robust regression: Asymptotics, conjectures, and Monte Carlo, Ann. Statist., 1, 799-821.

___ (1972), Robust statistics: A review, Ann. Math. Statist., 43, 1041-1067.

Hussain, A. (1969), A mixed model for regressions, Biometrika, 56, 322-336.

IMSL (1984), International Mathematical and Statistical Library, Inc. Reference Manual, Houston, Texas.

Izan, H. Y. (1980), To pool or not to pool? A reexamination of Tobin's food demand problem, J. Econometrics, 13, 391-402.

Jakubson, G. and Kiefer, N. (1985), Comment, Econometric Rev., 4, 175-178.

Jalilvand, A. and Harris, R. S. (1984), Corporate behavior in adjusting to capital structure and dividend targets: An econometric study, J. Finance, 39, 127-145.

Johansen, S. (1983), Some topics in regression, Scand. J. Statist., 10, 161-194.

___ (1982), Asymptotic inference in random coefficient regression models, Scand. J. Statist., 9, 201-207.

Johnson, J. A. and Oksanen, E. H. (1977), Estimation of demand for alcoholic beverages in Canada from pooled time series and cross sections, Rev. Econom. Statist., 59, 113-118.

___ (1974), Socio-economic determinants of the consumption of alcoholic beverages, Appl. Econom., 6, 292-301.

Johnson, K. H. and Lyon, H. L. (1973), Experimental evidence on combining cross-section and time series information, Rev. Econom. Statist., 55, 465-474.

Johnson, L. W. (1980a), Regional demand for gasoline: Comment, J. Regional Sci., 20, 99-101.

___ (1980b), Stochastic parameter regression: An additional annotated bibliography, Internat. Statist. Rev., 48, 95-102.

___ (1978), Regression with random coefficients, OMEGA, 6, 71-81.

___ (1977), Stochastic parameter regression: An annotated bibliography, Internat. Statist. Rev., 45, 257-272.

___ (1975), A note on testing for intraregional economic homogeneity, J. Regional Sci., 15, 365-369 (with comments and reply, see Vol. 17, 1977, pp. 473-479).

Johnson, L. and Hensher, D. (1982), Application of multinomial probit to a two-period panel data set, Transpn. Res., 16A, 457-464.

Johnson, L. W. and Oakenfull, E. A. (1978), RANPAR: A computer program for estimation of a random coefficient model with pooled time series and cross-section data, J. Marketing Res., 15, 611-612.

Johnson, P. R. (1964), Some aspects of estimating statistical cost functions, J. Farm. Econom., 46, 179-187.

___ (1960), Land substitutes and changes in corn yields, J. Farm. Econom., 42, 294-306.

Johnston, J. (1972), Econometric Methods, New York: McGraw-Hill Book Company.

Joreskog, K. G. (1978), An econometric model for multivariate panel data, Ann. I.N.S.E.E., 30-31, 355-366.

Joreskog, K. and Sorbom, D. (1978), Analysis of linear structural relationships by the method of maximum likelihood, LISREL IV, Release 2, November.

Judge, G. and Bock, M. (1978), *The Statistical Implications of Pre-Test and Stein-Rule Estimators in Econometrics*, Amsterdam: North-Holland Publishing Co.

Judge, G. G., Griffiths, W. E., Hill, R. C., Lutkepohl, H., and Lee, T. (1985), *The Theory and Practice of Econometrics*, New York: John Wiley and Sons, 2nd Edition.

Kadiyala, K. R. and Oberhelman, D. (1982), Response predictions in regressions on panel data, *Comm. Statist. A—Theory Methods*, 11, 2699-2714.

Kakwani, N. C. (1974), A note on the efficiency of Zellner's seemingly unrelated regressions estimator, *Ann. Inst. Statist. Math.*, 26, 361-362.

——— (1968), Note on the unbiasedness of a mixed regression estimator, *Econometrica*, 36, 610-611.

——— (1967), The unbiasedness of Zellner's seemingly unrelated regression equation estimators, *J. Amer. Statist. Assoc.*, 62, 141-142.

Kang, S. (1983), *Estimation of an Almost Ideal Demand System from Panel Data*, Unpublished Ph.D. Dissertation, University of Wisconsin-Madison.

Karathanassis, G. and Tzoannos, J. (1977), The demand for money by business firms: A temporal and cross-sectional analysis, *Appl. Econom.*, 9, 63-76.

Kariya, T. (1981a), Bounds for the covariance matrices of Zellner's estimator in the SUR model and the 2SAE in a heteroskedastic model, *J. Amer. Statist, Assoc.*, 76, 975-979.

——— (1981b), Tests for independence between two seemingly unrelated regression equations, *Ann. Statist.*, 9, 381-390.

Kariya, T. and Maekawa, K. (1982), A method for approximations to the pdf's and cdf's of GLSE's and its application to the seemingly unrelated regression model, *Ann. Inst. Statist. Math.*, 34, 281-297.

Kataoka, Y. (1974), The exact finite sample distribution of joint least squares estimators for seemingly unrelated regression equations, *Econom. Stud. Quart.*, 25, 36-44.

Kelejian, H. H. and Stephan, S. W. (1983), Inference in random coefficient panel data models: A correction and clarification of the literature, Internat. Econom. Rev., 24, 249-254.

Kiefer, N. M. (1980), estimation of fixed effect models for time series of cross-sections with arbitrary intertemporal covariance, J. Econometrics, 14, 195-202.

___ (1979), Population heterogeneity and inference from panel data on the effects of vocational education, J. Political Econom., 87, S213-S226.

Klein, L. R. (1953), A Textbook of Econometrics, Evanston: Row Peterson and Co.

Kmenta, J. (1971), Elements of Econometrics, New York: MacMillan Publishing Co., Inc.

Kmenta, J. and Gilbert, R. (1970), Estimation of seemingly unrelated regressions with autoregressive disturbances, J. Amer. Statist. Assoc., 65, 186-197.

___ (1968), Small sample properties of alternative estimators of seemingly unrelated regressions, J. Amer. Statist. Assoc., 65, 1180-1200.

Koenker, R. W. (1982), Robust methods in econometrics, Econometric Rev., 1, 213-290.

Kraft, J. and Rodekohr, M. (1980), Regional demand for gasoline: A reply to some considerations, J. Regional Sci., 20, 111-114.

___ (1978), Regional demand for gasoline: A temporal cross-section specification, J. Regional Sci., 18, 45-55 (with comments and reply, see Vol. 20, 1980, pp. 99-114).

Kuh, E. (1974), An essay on aggregation theory and practice, in Econometrics and Economic Theory (Ed. W. Sellekaerts). White Plains, N.Y.: International Arts and Sciences Press, Inc., 57-99.

___ (1959), The validity of cross-sectionally estimated behavior equations in time series applications, Econometrica, 27, 197-214.

Kuh, E. and Meyer, J. R. (1957), How extraneous are extraneous estimates? Rev. Econom. Statist., 39, 380-393.

Kunitomo, N. (1977), A note on the efficiency of Zellner's estimator for the case of two seemingly unrelated regression equations, Econom. Stud. Quart., 28, 73-77.

Kwast, M. L. (1980), A note on the structural stability of gasoline demand and the welfare economics of gasoline taxation, Southern Econom. J., 46, 1212-1220.

Laird, N. M. and Ware, J. H. (1982), Random-effects models for longitudinal data, Biometrics, 38, 963-974.

Langetieg, T. C. (1978), An application of a three factor performance index to measure stockholder gains from merger, J. Finan. Econ., 6, 365-384.

Langetieg, T. C., Haugen, R. and Wichern, D. (1980), Merger and stockholder risk, J. Finan. Quant. Analysis, 15, 689-718.

Lee, C. (1976), A note on the interdependent structure of security returns, J. Fin. Quant. Analysis, 11, 73-86.

Lee, C. F. and Chang, H. (1986), Dividend policy and capital market theory: A generalized error-components model approach, J. Business Res., 14, 177-188.

Leonard, J. S. (1982), Wage expectations in the labor market: Survey evidence on rationality, Rev. Econom. Statist., 64, 157-161.

Li, W. K. and Hui, Y. V. (1983), Estimation of random coefficient autoregressive processes: An empirical Bayes approach, J. Time Series Analysis, 4, 89-94.

Lillard, L. A. (1978), Estimation of permanent and transitory response functions in panel data: A dynamic labor supply model, Ann. I.N.S.E.E., 30-31, 367-394.

Lillard, L. and Weiss, Y. (1979), Components of variation in panel earnings data: American scientists 1960-70, Econometrica, 47, 437-454.

Lin, A. (1985), A note on testing for regional homogeneity of a parameter, J. Regional Sci., 25, 129-134.

Liu, L. and Hanssens, D. M. (1981), A Bayesian Approach to time-varying cross-sectional regression models, J. Econometrics, 15, 341-356.

Liu, L. and Tiao, G. C. (1980), Random coefficient first-order autoregressive models, J. Econometrics, 13, 305-325.

MaCurdy, T. E. (1982), The use of time series processes to model the error structure of earnings in a longitudinal data analysis, J. Econometrics, 18, 83-114.

Maddala, G. S. (1978), Selectivity problems in longitudinal data, Ann. I.N.S.E.E., 30-31, 423-450.

___ (1977), Econometrics, New York: McGraw Hill Book Co.

___ (1971a), The likelihood approach to pooling cross-section and time series data, Econometrica, 39, 939-953.

___ (1971b), The use of variance components models in pooling cross section and time series data, Econometrica, 39, 341-358.

Maddala, G. S. and Mount, T. D. (1973), A comparative study of alternative estimators for variance components models used in econometric applications, J. Amer. Statist. Assoc., 68, 324-328.

Maekawa, K. (1983), Effects of non-orthogonal regressors on the distribution of seemingly unrelated regression estimator, J. Japan. Statist. Soc., 13, 145-149.

Maeshiro, A. (1980), New evidence on the small sample properties of estimators of SUR models with autocorrelated disturbances: Things done halfway may not be done right, J. Econometrics, 12, 177-187.

___ (1979), On the retention of the first obervation in serial correlation adjustment of regression models, Internat. Econom. Rev., 20, 487-489.

Magnus, J. R. (1982), Multivariate error components analysis of linear and nonlinear regression models by maximum likelihood, J. Econometrics, 19, 239-285.

___ (1978), Maximum likelihood estimation of the GLS model with unknown parameters in the disturbance covariance matrix, J. Econometrics, 7, 281-312.

Mairesse, J. (1978), New estimates of embodied and disembodied technical progress, Ann. I.N.S.E.E., 30-31, 681-720.

Mandelker, G. (1974), Risk and return: The case of merging firms, J. Finan. Econom., 1, 303-335.

Manski, C. and McFadden, D. (1981), Alternative estimators and sample designs for discrete choice analysis, in Structural Analysis of Discrete Data with Econometric Applications, (Eds. C. Manski and D. McFadden) Cambridge: MIT Press, 2-50.

Marcis, R. G. and Smith, V. K. (1974), Efficient estimation of multivariate financial relationships, J. Finance, 29, 1415-1423.

___ (1973), The demand for liquid asset balances by U. S. manufacturing corporations: 1959-1970, J. Fin. Quant. Analysis, 8, 207-218.

Margolis, S. E. (1982), Depreciation of housing: An empirical consideration of the filtering hypothesis, Rev. Econom. Statist., 64, 90-96.

Marschak, J. (1943), Money illusion and demand analysis, Rev. Econom. Statist., 25, 40-48.

Mazodier, P. and Trognon, A. (1978), Heteroscedasticity and stratification in error components models, Ann. I.N.S.E.E., 30, 451-482.

McElroy, M. (1977a), Goodness of fit for seemingly unrelated regressions, J. Econometrics, 6, 381-387.

___ (1977b), Weaker MSE criteria and tests for linear restrictions in regression models with non-spherical disturbances, J. Econometrics, 6, 389-394.

McFadden, D. (1984), Econometric analysis of qualitative response models, in Handbook of Econometrics, Vol. 2 (Eds. Z. Griliches and M. Intriligator) Amsterdam: North-Holland, 1395-1457.

___ (1982), Qualitative response models, in Advances in Econometrics, Proceedings of World Meetings of the Econometric Society, 1980 (Ed. W. Hildenbrand) Cambridge: Cambridge University Press, 1-37.

___ (1974), Conditional logit analysis of qualitative choice behavior, in Frontiers in Econometrics, (Ed. P. Zarembka) New York: Academic Press, 105-142.

Mehta, J. S. and Swamy, P.A.V.B. (1976), Further evidence on the relative efficiencies of Zellner's seemingly unrelated regressions estimator, J. Amer. Statist. Assoc., 71, 634-639.

Mehta, J. S. and Swamy, P.A.V.B. (1974), The exact finite sample distribution of Theil's compatibility test statistic and its application, J. Amer. Statist. Asso., 69, 154-158.

___ (1970), The finite sample distribution of Theil's mixed regression estimator and a related problem, Rev. Internat. Statist. Inst., 38, 202-209.

Mehta, J. S., Narasimham, G. V. L., and Swamy, P.A.V.B. (1978), Estimation of a dynamic demand function for gasoline with different schemes of parameter variation, J. Econometrics, 7, 263-279.

Mikhail, W. M. (1975), Pooling time-series and cross-section data when the time-series equation is one of a complete system, Egyptian Statist. J., 57-65.

Miller, M. K. and Voth, D. E. (1982), Rural health service programs: Evaluating with pooled cross-section and time series data, Evaluation Rev., 6, 47-59.

Moffitt, R. (1984), Profiles of fertility, labour supply and wages of married women: A complete life-cycle model, Rev. Econom. Studies, 61, 263-278.

Moriarty, M. (1975), Cross-sectional, time-series issues in the analysis of marketing decision variables, J. Marketing Res., 12, 142-150.

Mundlak, Y. (1978a), On the pooling of time series and cross section data, Econometrica, 46, 69-85.

___ (1978b), Models with variable coefficients: Integration and extension, Ann. I.N.S.E.E., 30-31, 483-509.

___ (1963), Estimation of production and behavioral functions from a combination of cross-section and time-series data, in Measurement in Economics, ed., C. F. Christ el al., Stanford, Ca: Stanford University Press, 138-166.

___ (1961), Empirical production functions free of management bias, J. Farm Econom., 43, 44-56.

Mundlak, Y. and Yahav, J. A. (1981), Random effects, fixed effects, convolution, and separation, Econometrica, 49, 1399-1416.

Nagar, A. L. and Kakwani, N. C. (1964), The bias and moment matrix of a mixed regression estimator, Econometrica, 32, 174-182.

Nelson, R. A. and Wohar, M. E. (1983), Regulation, scale economics, and productivity in steam-electric generation, Internat. Econom. Rev., 24, 57-79.

Nerlove, M. (1971a), A note on error components models, Econometrica, 39, 383-396.

___ (1971b), Further evidence on the estimation of dynamic economic relations from a time series of cross sections, Econometrica, 39, 359-382.

___ (1967), Experimental evidence on the estimation of dynamic economic relations from a time series of cross-sections, Econom. Stud. Quart., 18, 42-74.

___ (1965), Estimation and Identification of Cobb-Douglas Production Functions, Chicago: Rand McNally and Co.

Nickell, S. (1981), Biases in dynamic models with fixed effects, Econometrica, 49, 1417-1426.

Nielsen, F., and Hannan, M. (1977), The expansion of national educational systems: Tests of a population ecology model, Amer. Sociological Rev., 42, 479-490.

Norberg, R. (1977), Inference in random coefficient regression models with one-way and nested classifications, Scand. J. Statist., 4, 71-80.

Nssah, B. E. (1982), Pooling possibly heterogeneous data, Unpublished Ph.D. Dissertation, University of Michigan.

Oberhofer, W. and Kmenta, J. (1974), A general procedure for obtaining maximum likelihood estimates in generalized regression models, Econometrica, 42, 579-590.

Ohtani, K. and Honda, Y. (1984), Small sample properties of the mixed regression estimator, J. Econometrics, 26, 375-385.

Oppenheimer, H. R. and Dielman, T. E. (1988), Firm dividend policy and insider activity: Some empirical results, J. Business Fin. Accounting, forthcoming.

Oudiz, G. (1978), Investment behavior of French industrial firms: A study on longitudinal data, Ann. I.N.S.E.E., 30-31, 511-541.

Owusu-Gyapong, A. (1986), Alternative estimating techniques for panel data on strike activity, Rev. Econom. Statist., 68, 526-531.

Pakes, A. and Griliches, Z. (1984), Estimating distributed lags in short panels with an application to the specification of depreciation patterns and capital stock constructs, Rev. Econom. Studies, 51, 243-262.

Palda, K. S. and Blair, L. M. (1970), A moving cross-section analysis of demand for tooth paste, J. Marketing Res., 7, 439-449.

Parhizgari, A. and Davis, P. (1978), The residential demand for electricity: A variant parameters approach, Appl. Econom., 10, 331-340.

Park, R. E. and Mitchell, B. M. (1980), Estimating the autocorrelated error model with trended data, J. Econometrics, 13, 185-201.

Park, R. E., Mitchell, B. M., Wetzel, B. M., and Alleman, J. H. (1983), Charging for local telephone calls: How household characteristics affect the distribution of calls in the GTE Illinois experiment, J. Econometrics, 22, 339-364.

Parks, R. (1967), Efficient estimation of a system of regression equations when disturbances are both serially and contemporaneously correlated, J. Amer. Statist. Assoc., 62, 500-509.

Parsons, L. (1974), An econometric analysis of advertising, retail availability, and sales of a new brand, Management Sci., 20, 938-947.

Paulus, J. D. (1975), Mixed estimation of a complete system of consumer demand equations, Ann. Econom. Social Measurement, 4, 117-131.

Phillips, P. C. B. (1985), The exact distribution of the SUR estimator, Econometrica, 53, 745-756.

Phillips, P. C. B. (1977), An approximation to the finite sample distribution of Zellner's seemingly unrelated regression estimator, J. Econometrics, 6, 147-164.

Pindyck, R. and Rubinfeld, D. (1976), Econometric Models and Economic Forecasts, New York: McGraw-Hill Book Co.

Porter, R. D. (1973), On the use of survey sample weights in the linear model, Ann. Econom. Social Measurement, 2, 141-158.

Prucha, I. R. (1985), Maximum likelihood and instrumental variable estimation in simultaneous equation systems with error components, Internat. Econom. Rev., 26, 491-506.

___ (1984), On the asymptotic efficiency of feasible Aitken estimators for seemingly unrelated regression models with error components, Econometrica, 52, 203-207.

Pudney, S. E. (1978), The estimation and testing of some error components models, London School of Economics, mimeo.

Ramanathan, R. and Mitchem, A. (1982), Econometric and computational issues in estimating demand for energy by time of day, Rev. Econom. Statist., 64, 335-339.

Rao, C. R. (1972), Estimation of variance and covariance components in linear models, J. Amer. Statist. Assoc., 67, 112-115.

___ (1971a), Estimation of variance and covariance components—MINQUE theory, J. Multivariate Anal., 1, 257-275.

___ (1971b), Minimum variance quadratic estimation of variance components, J. Multivariate Anal., 1, 445-456.

___ (1970), Estimation of heteroscedastic variances in linear models, J. Amer. Statis. Assoc., 65, 161-172.

___ (1965), The theory of least squares when the parameters are stochastic and its application to the analysis of growth curves, Biometrika, 52, 447-458.

Rao, C. R. and Mitra, S. K. (1971), Generalized Inverse of Matrices and its Applications, New York: John Wiley and Sons, Inc.

Rao, P. (1974), Specification bias in seemingly unrelated regressions, in Econometrics and Economic Theory (Ed. W. Sellekaerts), White Plains, N. Y.: International Arts and Sciences Press, Inc., 101-113.

Rao, U. L. G. (1982), A note on the unbiasedness of Swamy's estimator for the random coefficient regression model, J. Econometrics, 18, 395-401.

Rausser, C. G. and Oliver, R. O. (1976), An econometric analysis of wilderness area use, J. Amer. Statist. Assoc., 71, 276-285.

Rayner, R. K. (1981), The prediction of systematic risk using equivalent risk classes, Unpublished Working Paper No. 252, Graduate School of Business Administration, University of Michigan.

Rayner, R. K. and Wright, R. L. (1980), Efficient inference in random coefficient models with multicollinearity in the time series regressions, Amer. Statist. Assoc. Business Econom. Statist. Proc., 59-64.

Reinsel, G. C. (1985), Mean squared error properties of empirical Bayes estimators in a multivariate random effects general linear model, J. Amer. Statist. Assoc., 80, 642-650.

___ (1984), Estimation and prediction in a multivariate random effects generalized linear model, J. Amer. Statist. Assoc., 79, 406-414.

Render, M. W. and Neumann, G. R. (1980), Conflict and contrast: The case of strikes, J. Political Econom., 88, 867-886.

Revankar, N. (1976), Use of restricted residuals in SUR systems: Some finite sample results, J. Amer. Statist. Assoc., 71, 183-188.

___ (1974), Some finite sample results in the context of two seemingly unrelated regression equations, J. Amer. Statist. Assoc., 69, 187-190.

Rodekohr, M. (1979), Demand for transportation fuels in the OECD: A temporal cross-section specification, Appl. Energy, 5, 223-231.

Rosenberg, B. (1974), Extra-market components of covariance in security returns, J. Fin. Quant. Analysis, 9, 263-274.

___ (1973a), A survey of stochastic parameter regression, Ann. Econom. Social Measurement, 2, 381-397.

___ (1973b), The analysis of a cross section of time series by stochastically convergent parameter regression, Ann. Econom. Social Measurement, 2, 399-428.

___ (1973c), Linear regression with randomly dispersed parameters, Biometrika, 60, 65-72.

Rosenberg, B. and McKibben, W. (1973), The prediction of systematic and specific risk in common stocks, J. Fin. Quant. Analysis, 8, 317-333.

Rosenberg, S. H. and Levy, P. S. (1972), A characterization on missspecification in the general linear regression model, Biometrics, 28, 1129-1132.

Rothenberg, T. J. (1984), Approximate normality of generalized least squares estimates, Econometrica, 52, 811-825.

Roy, S. N. (1957), Some Aspects of Multivariate Analysis, New York: John Wiley and Sons, Inc.

Ryan, B., Joiner, B., and Ryan, T. (1985), Minitab Handbook, Boston: Duxbury Press, 2nd Ed.

Salvas-Bronsard, L. (1978), Estimating systems of demand equations from French time-series of cross-sections data, Ann. I.N.S.E.E., 30-31, 543-564.

SAS User's Guide: Basics (1982), Cary, North Carolina: SAS Institute, Inc.

Schmalensee, R. (1976), An experimental study of expectation formation, Econometrica, 44, 17-41.

___ (1972), Variance estimation in a random coefficient regression model, Discussion Paper 72-10, Department of Economics, University of California, San Diego.

Schmidt, P. (1983), A note on a fixed effect model with arbitrary interpersonal covariance, J. Econometrics, 22, 391-393.

___ (1978), A note on the estimation of seemingly unrelated regression systems, J. Econometrics, 7, 259-261.

___ (1977), Estimation of seemingly unrelated regressions with unequal numbers of observations, J. Econometrics, 5, 365-377.

Schmidt, P. and Sickles, R. C. (1984), Production frontiers and panel data, J. Business Econom. Statist., 2, 367-374.

Schwartz, R. W. and Whitcomb, D. K. (1977), Evidence on the presence and causes of serial correlation in market model residuals, J. Finan. Quant. Analysis, 32, 291-313.

Shapiro, D. and Mott, E. F. (1978), Labor force attachment during the early childbearing years: Evidence from the national longitudinal surveys of young women, Ann. I.N.S.E.E., 30-31, 565-598.

Sheiner, L., Rosenberg, B. and Melmon, K. (1972), Modeling of individual pharmacokinetics for computer-aided drug dosage, Comput. Biomedical Res., 5, 441-459.

Sickles, R. C. (1985), A nonlinear multivariate error components analysis of technology and specific factor productivity growth with an application to the U.S. airlines, J. Econometrics, 27, 61-78.

Silver, J. L. (1982), Generalized estimation of error components models with a serially correlated temporal effect, Internat. Econom. Rev., 23, 463-478.

Singer, B. (1981), Estimation of nonstationary Markov chains from panel data, in Sociological Methodology, (Ed. S. Leinhardt), San Francisco: Jossey-Bass Inc., Publishers, 319-337.

Singer, B. and Cohen, J. E. (1980), Estimating malaria incidence and recovery rates from panel surveys, Math Biosci., 49, 273-305.

Singer, B. and Spilerman, S. (1976), Some methodological issues in the analysis of longitudinal surveys, Ann. Econom. Social Measurement, 5, 447-474.

Singh, B. and Ullah, A. (1974), Estimation of seemingly unrelated regressions with random coefficients, J. Amer. Statist. Assoc., 69, 191-195.

Solow, R. M. (1964), Capital, labor and income in manufacturing, in Studies in Income and Wealth, 27, Princeton: NBER, Princeton University Press.

Spencer, D. (1979), Estimation of a dynamic system of seemingly unrelated regressions with autoregressive disturbances, J. Econometrics, 10, 227-241.

Spjotvoll, E. (1977), Random coefficient regression models, a review, Math. Operationsforsch. Statist. Ser. Statist., 8, 69-93.

SPSSX User's Guide (1986), New York: McGraw-Hill Book Co., 2nd Ed.

Srivastava, A. and Tracy, D. (1986), Computation of standard errors in seemingly unrelated regression equation models, Commun. Statist. - Theory Meth., 15, 3583-3588.

Srivastava, S. and Srivastava, V. (1983), Estimation of the seemingly unrelated regression model under specification error, Biomedical J., 25, 181-191.

Srivastava, V. K. (1973), The efficiency of an improved method of estimating seemingly unrelated regression equations, J. Econometrics, 1, 341-350.

___ (1970), The efficiency of estimating seemingly unrelated regression equations, Ann. Instit. Statist. Math., 22, 483-493.

Srivastava, V. K. and Dwivedi, T. D. (1979), Estimation of seemingly unrelated regression equations: A brief survey, J. Econometrics, 10, 15-32.

Srivastava, V. K. and Giles, D. E. A. (1987), Seemingly Unrelated Regression Equations Models: Estimation and Inference, New York: Marcel Dekker, Inc.

Srivastava, V. K. and Raj, B. (1979), The existence of the mean of the estimator in seemingly unrelated regressions, Comm. Statist. A - Theory Methods, 8, 713-717.

Srivastava, V. K. and Srivastava, A. K. (1984), Improved estimation in a two equation seemingly unrelated regression model, Statistica, 44, 417-422.

Srivastava, V. K. and Upadhyaya, S. (1978), Large-sample approximations in seemingly unrelated regression equations, Ann. Instit. Statist. Math., 30, 89-96.

Staehle, H. (1945), Relative prices and postwar markets for animal food products, Quart. J. Econom., 59, 237-279.

Stanek, E. J. and Koch, G. G. (1985), The equivalence of parameter estimates from growth curve models and seemingly unrelated regression models, Amer. Statistician, 39, 149-152.

Stokes, H. H. (1981), The B34S data analysis program: A short writeup, Report FY77-1, College of Business Administration Working Paper Series, University of Illinois at Chicago Circle.

Stone, J. R. N. (1954), The Measurement of Consumer's Expenditure and Behavior in the United Kingdom, 1928-1938, Cambridge: Cambridge University Press.

Suzuki, Y. (1964), On the use of some extraneous information in the estimation of coefficients of regression, Ann. Inst. Statist. Math., 16, 161-173.

Swamy, P. A. V. B. (1974), Linear models with random coefficients, in Frontiers in Econometrics (Ed. P. Zarembka), New York: Academic Press, Inc., 143-168.

___ (1973), Criteria, constraints and multicollinearity in random coefficient regression models, Ann. Econom. Social Measurement, 2, 429-450.

___ (1971), Statistical Inference in Random Coefficient Regression Models, Berlin: Springer-Verlag.

___ (1970), Efficient inference in a random coefficient regression model, Econometrica, 38, 311-323.

Swamy, P. A. V. B. and Arora, S. (1972), The exact finite sample properties of the estimators of coefficients in the error components regression models, Econometrica, 40, 261-275.

Swamy, P. A. V. B. and Mehta, J. S. (1979), Estimation of common coefficients in two regression equations, J. Econometrics, 10, 1-14.

___ (1977a), Estimation of linear models with time and cross-sectionally varying coefficients, J. Amer. Statist. Assoc., 72, 890-898.

___ (1977b), Robustness of Theil's mixed regression estimators, Canad. J. Statist., 5, 93-109.

___ (1975a), Bayesian and non-Bayesian analysis of switching regressions and of random coefficient regression models, J. Amer. Statist. Assoc., 70, 593-602.

___ (1975b), On Bayesian estimation of seemingly unrelated regressions when some observations are missing, J. Econometrics, 3, 157-169.

___ (1973a), Bayesian analysis of error components regression models, J. Amer. Statist. Assoc., 68, 648-658.

___ (1969), On Theil's mixed regression estimator, J. Amer. Statist. Assoc., 64, 273-276.

Swamy, P. A. V. B. and Tinsley, P. A. (1980), Linear prediction and estimation methods for regression models with stationary stochastic coefficients, J. Econometrics, 12, 103-142.

Taub, A. (1979), Prediction in the context of the variance-components model, J. Econometrics, 10, 103-107.

Taylor, W. E. (1980), Small sample considerations in estimation from panel data, J. Econometrics, 13, 203-223.

Telser, L. G. (1964), Iterative estimation of a set of linear regression equations, J. Amer. Statist. Assoc., 59, 845-862.

Terasvirta, T. (1975), A note on predicting with seemingly unrelated regression equations, Math. Operationsforsch. Statist. Ser. Statist., 6, 709-711.

Theil, H. (1974), Mixed estimation based on quasi-prior judgements, European Econom. Rev., 5, 33-40.

___ (1971), Principles of Econometrics, New York: John Wiley and Sons, Inc.

___ (1963), On the use of incomplete prior information in regression analysis, J. Amer. Statist. Assoc., 58, 401-414.

___ (1954), Linear Aggregation of Economic Relations, Amsterdam: North-Holland Publishing Co.

Theil, H. and Goldberger, A. S. (1961), On pure and mixed statistical estimation in economics, Internat. Econom. Rev., 2, 65-78.

Tiao, G. C., Tan, W. and Chang, Y. (1977), Some aspects of bivariate regression subject to linear constraints, J. Econometrics, 5, 13-35.

Tobin (1950), A statistical demand function for food in the U.S.A., J. Roy. Statist. Soc., Ser. A, 113, 113-141.

Toro-Vizcarrondo, C. and Wallace, T. D. (1968), A test of the mean square error criterion for restrictions in linear regression, J. Amer. Statist. Assoc., 63, 558-572.

Trivedi, P. K. (1980), Small samples and collateral information: An application of the hyperparameter model, J. Econometrics, 12, 301-318.

Trognon, A. (1978), Miscellaneous asymptotic properties of ordinary least squares and maximum likelihood estimators in dynamic error components models, Ann. I.N.S.E.E., 30-31, 631-657.

Verbon, H. A. A. (1980) Maximum likelihood estimation of a labour demand system. An application of a model of seemingly unrelated regression equations with the regression errors composed of two components, Statist. Neerlandica, 34, 33-48.

Vernon, J. M. and McElroy, M. B. (1973), Estimation of structure-profit relationships: Comment, Amer. Econom. Rev., 63, 763-767.

Vinod, H. and Ullah, A. (1981), Recent Advances in Regression Methods, New York: Marcel Dekker, Inc.

Wallace, T. D. (1972), Weaker criteria and tests for linear restrictions in regression, Econometrica, 40, 689-698.

Wallace, T. D., and Hussain, A. (1969), The use of error components models in combining cross section with time series data, Econometrica, 37, 55-72.

Wallace, T. D. and Toro-Vizcarrondo, C. E. (1969), Tables for the mean square error test for exact linear restrictions in regression, J. Amer. Statist. Assoc., 64, 1649-1663.

Wang, G. H. K., Hidiroglou, M., and Fuller, W. A. (1980), Estimation of seemingly unrelated regression with lagged dependent variables and autocorrelated errors, J. Statist. Computation Simulation, 10, 133-146.

Wansbeek, T. and Kapteyn, A. (1982), A class of decompositions of the variance-covariance matrix of a generalized error components model, Econometrica, 50, 713-724.

___ (1978), The separation of individual variation and systematic change in the analysis of panel data, Ann. I.N.S.E.E., 30-31, 659-680.

Ward, R. and Davis, J. (1978), A pooled cross-section time series model of coupon promotions, Amer. J. Agricultural Econom., 60, 393-401.

Ware, J. H. (1985), Linear models for the analysis of longitudinal studies, Amer. Statistician, 39, 95-101.

Weiss, Y. and Lillard, L. (1978), Experience, vintage, and time effects in the growth of earnings: American scientists, 1960-1970. J. Political Econom., 86, 427-447.

Welsch, R., and Kuh, E. (1976), The variances of regression coefficient estimates using aggregate data, Econometrica, 44, 353-363.

White, K. J. (1978), A general computer program for econometric methods--SHAZAM, Econometrica, 46, 239-240.

Wilbur, W. L., Miller, G. L., and Brown, W. J. (1985), A pooled time-series-cross-section analysis of returns on alternative sources of bank capital, Appl. Econom., 17, 1023-1041.

Wildt, A. (1974), Multifirm analysis of competitive decision variables, J. Marketing Res., 11, 50-62.

Wilson, J. W. (1974), Electricity consumption: Supply requirements, demand elasticity and rate design, Amer. J. Agricultural Econom., 56, 419-426 (with discussion, 427-435).

Wittink, D. (1977), Exploring territorial differences in the relationship between marketing variables, J. Marketing Res., 14, 145-155.

Wold, H. and Jureen, L. (1951), Demand Analysis, New York: John Wiley and Sons, Inc.

Wolpin, K. I. (1980), A time series-cross section analysis of international variation in crime and punishment, Rev. Econom. Statistics, 62, 417-423.

Woodland, L. (1983), Development of a Cross-Sectionally Constrained Random Coefficient Model, with an Application to the Estimation of the Vintage Effect on Earnings Profiles, Unpublished Ph.D. Dissertation, Purdue University.

Yancey, T. A., Bock, M. E., and Judge, G. G. (1972), Some finite sample results from Theil's mixed regression estimator, J. Amer. Statist. Assoc., 67, 176-179.

Young, K. H. (1972), A synthesis of time-series and cross-section analyses: Demand for air transportation service, J. Amer. Statist. Assoc., 67, 560-566.

Zellner, A. (1971), An Introduction to Bayesian Inference in Econometrics, New York: John Wiley and Sons.

___ (1969), On the aggregation problem: A new approach to a troublesome problem, in Economic Models, Estimation and Risk Programming: Essays in Honor of Gerhard Tintner (Eds. Fox, K. A. et al.), New York: Springer-Verlag, 365-374.

___ (1963), Estimators for seemingly unrelated regression equations: Some exact finite sample results, J. Amer. Statist. Assoc., 58, 977-992. (See also Corrigenda in March 1972 JASA, 255).

___ (1962), An efficient method of estimating seemingly unrelated regressions and tests for aggregation bias, J. Amer. Statist. Assoc., 57, 348-368.

Zellner, A. and Huang, D. S. (1962), Further properties of efficient estimators for seemingly unrelated regression equations, Internat. Econom. Rev., 3, 300-313.

Zellner, A., and Vandaele, W. (1975), Bayes-Stein estimators for k-means, regression and simultaneous equation models, in Studies in Bayesian Econometrics and Statistics, (Ed. S. E. Fienberg and A. Zellner), Amsterdam: North Holland, 627-653.

Index

G

K

L

M

O

P